DHARMA AND ECOLOGY OF HINDU COMMUNITIES

In Indic religious traditions, a number of rituals and myths exist in which the environment is revered. Despite this nature worship in India, its natural resources are under heavy pressure with its growing economy and exploding population. This has led several scholars to raise questions about the role religious communities can play in environmentalism. Does nature worship inspire Hindus to act in an environmentally conscious way? This book explores the above questions with three communities, the Swadhyaya movement, the Bishnoi, and the Bhil communities. Presenting the texts of Bishnois, their environmental history, and their contemporary activism; investigating the Swadhyaya movement from an ecological perspective; and exploring the Bhil communities and their Sacred Groves, this book applies a non-Western hermeneutical model to interpret the religious traditions of Indic communities.

ASHGATE NEW CRITICAL THINKING IN RELIGION, THEOLOGY AND BIBLICAL STUDIES

The *Ashgate New Critical Thinking in Religion, Theology and Biblical Studies* series brings high quality research monograph publishing back into focus for authors, international libraries, and student, academic and research readers. Headed by an international editorial advisory board of acclaimed scholars spanning the breadth of religious studies, theology and biblical studies, this open-ended monograph series presents cutting-edge research from both established and new authors in the field. With specialist focus yet clear contextual presentation of contemporary research, books in the series take research into important new directions and open the field to new critical debate within the discipline, in areas of related study, and in key areas for contemporary society.

Other Recently Published Titles in the Series:

What's Right with the Trinity?
Conversations in Feminist Theology
Hannah Bacon

Spirit and Sonship
Colin Gunton's Theology of Particularity and the Holy Spirit
David A. Höhne

Dalit Theology and Dalit Liberation
Problems, Paradigms and Possibilities
Peniel Rajkumar

Beyond Evangelicalism
The Theological Methodology of Stanley J. Grenz
Steven Knowles

Concepts of Power in Kierkegaard and Nietzsche
J. Keith Hyde

Kierkegaard, Pietism and Holiness
Christopher B. Barnett

The Trinity and Theodicy
The Trinitarian Theology of von Balthasar and the Problem of Evil
Jacob H. Friesenhahn

Dharma and Ecology of Hindu Communities

Sustenance and Sustainability

PANKAJ JAIN
University of North Texas, USA

ASHGATE

Published by

Ashgate Publishing Limited
Wey Court East
Union Road
Farnham
Surrey, GU9 7PT
England

Ashgate Publishing Company
Suite 420
101 Cherry Street
Burlington
VT 05401-4405
USA

www.ashgate.com

British Library Cataloguing in Publication Data
Jain, Pankaj.
Dharma and ecology of Hindu communities : sustenance and sustainability. -- (Ashgate new critical thinking in religion, theology and biblical studies)
1. Environmentalism--Religious aspects--Hinduism.
2. Human ecology--Religious aspects--Hinduism. 3. Dharma.
4. Swadhyaya Parivar (Organization) 5. Bishnois. 6. Bhil (Indic people) 7. India--Environmental conditions.
I. Title II. Series
294.5'177-dc22

Library of Congress Cataloging-in-Publication Data
Jain, Pankaj.
Dharma and ecology of Hindu communities : sustenance and sustainability / Pankaj Jain.
p. cm. -- (New critical thinking in religion, theology and biblical studies)
Includes index.
ISBN 978-1-4094-0591-7 (hardcover) -- ISBN 978-1-4094-0592-4 (ebook) 1. Ecology--Religious aspects--Hinduism. 2. Nature--Religious aspects--Hinduism. 3. Environmentalism--Religious aspects--Hinduism. 4. Hinduism--Customs and practices. 5. Natural history--India. 6. Sustainable development--India. 7. India--Religious life and customs. 8. India--Social life and customs. I. Title.
BL1215.N34J35 2010
294.5'177--dc22

2010046208

ISBN 9781409405917 (hbk)
ISBN 9781409405924 (ebk)

MIX
Paper from responsible sources
FSC® C018575
FSC www.fsc.org

Printed and bound in Great Britain by the MPG Books Group, UK

Contents

List of Figures vii
Foreword ix
Acknowledgments xi
List of Abbreviations xiii

1 Introduction 1

2 Theoretical Propositions for Indic Traditions and Ecology 5

3 The Swadhyaya Movement 17

4 The Bishnoi Community 51

5 Sacred Groves of Bhils 79

6 Modern Organizations Adapting to Ecology 95

7 Dharma as Religious and Environmental Ethos 105

8 Conclusions 117

APPENDICES

A Translation of Jambheśvara's Śabdas 135
B Hindu Myths in Jambheśvara's Śabdas 165
C Bishnoi Saṃskāras ("Rites of Passage") 171
D Translation of Jambheśvara Darśana by Brahmanand Sharma 179
E Athavale's Ecological Inspirations 181
F History of Beneśvara, a Bhil Pilgrimage Center 193

Bibliography 195
Index 211

List of Figures

4.1	The image of Jambheśvara inside a Bishnoi temple	54
4.2	A Bishnoi temple near Pali	62
4.3	A portrayal of massacre at Khejadali Memorial	64
4.4	The Sacrifice Memorial at Khejadali	66
4.5	Bishnoi farmers with khejari trees in their farm	67
5.1	View from inside a Bhil temple	87
5.2	A Hanumān temple inside a Bhil sacred grove	87
5.3	A Durgā temple inside a Bhil sacred grove	88
5.4	A sign at the Sītāmātā Sanctuary showing the birthplace of Sītā's two sons	91
5.5	Temple at Ghoṭiyā Ambā	92

Foreword

Roger S. Gottlieb[1]

In the face of the environmental crisis, we need all the help we can get. And, as this is perhaps the greatest crisis human civilization has ever faced, virtually all forms of contemporary culture, economics, politics, and spirituality will have to make profound changes. Yet who is the "we" who need help? And what can "we" turn to and rely on if all that situates, sustains, and directs us now is called into question?

It is out of such dialectical tensions that Pankaj Jain's fine book is born. On the one hand, Jain is clearly motivated by environmental awareness and concern. This is not a book which simply follows on some long established academic track. Jain is studying – and offering us valuable insights – because of a practical concern with the survival of life on earth. This is "organic" scholarship as Gramsci intended: learning oriented to the problems of living communities.

At the same time Jain's inquiry into Indian sources of ecological awareness and practice grounds "our" response to environmental crisis in the living practices and beliefs of particular communities. As universal as global warming, pollution, species extinction, deforestation, etc. are, we do not face them as "universal" beings, but as people situated by history, culture, and social position. In that particularity lies a variety of both resources and failures. One group will necessarily be more technologically sophisticated, but perhaps at the same time more addicted to consumption. One will feel itself more rooted in the earth, but lack the social capital to defend itself against the developers. Some will embrace an environmentally benign spirituality but shy away from spiritual social activism. And some may mistakenly separate the ethical concern with human beings from respect for other forms of and supports for life. Traditional religion may have much to teach us about how to respect or even revere nature. But now that nature – powerful, infinite, the source of all our lives – has become the fragile and often poisoned environment, significant shifts of religious understanding and practice are needed. Each group, community, and tradition will have its gifts, and its limitations.

The focused particularity of Jain's study, as well as his broad learning and sympathetic approach to his subjects, make this book particularly valuable. What do we in other countries have to learn from these Indian communities?

---

1 Roger S. Gottlieb is Professor of Philosophy at Worcester Polytechnic Institute. gottlieb@wpi.edu.

And what might we teach them? It is out of such intellectual, political, and human encounters that the global environmental movement is sustained, and I am grateful to Jain for his contribution to it.

Acknowledgments

This book introduces the ecological practices of three communities in India and the Indian diaspora. I have avoided the word caste or jati to refer to these communities because all three groups are bound together not so much by these notions as by their dharmic traditions preserved either in oral transmission (as in the case of Bhils), written scriptures in vernacular languages (as in the case of Bishnois), or recorded in video cassettes and books (as in the case of Swadhyayis).

I have correlated ecology with dharma in both the title of the book and in various places in its chapters. The terms sustenance and sustainability derive from the English verb "to sustain"; but their ideological cognates in India derive from the Sanskrit verbal root "dhri," from which the word "dharma" also derives. Thus, my attempt in this book is to explore both sustenance and sustainability from the dharma as lived by Swadhyayis, Bishnois, and Bhils.

Although this book began its formal life as a PhD dissertation, it is really an outcome of more than a decade of my friendship with Swadhyayis (participants of the Swadhyaya movement), whom I first met in the late 1990s in New Jersey and then continued meeting in North Carolina, Texas, and during my trips to India. I have continued to learn about the Swadhyaya activities, either as a participant, or as an observer, and often in my hyphenated identity as a participant-observer. My research about the Bishnois and the Bhils is based on my visits at their sacred places in Rajasthan in 2006 and later via electronic communication with some of the Bishnois.

After the successful completion of the book, an even more pleasant task is to express my gratitude to all the people who directly or indirectly helped during the project. I must start with my parents who planted and nurtured the seed of intellectual curiosity that finally yielded its fruit when I left my career as a software engineer to pursue graduate education in Religious Studies. I thank all the teachers of Indic traditions from whom I have benefited intellectually and spiritually including Reverend Pandurang Shastri Athavale, the founder of the Swadhyaya Parivar, Guru Jambheshwara, the founder of the Bishnoi tradition, and various gurus within the Bhil traditions. I would also like to express my deep gratitude to my professors at the University of Iowa and at Columbia University and other academic advisors for all the intellectual engagements over the years: Frederick M. Smith, Christopher K. Chapple, George A. James, Paul R. Greenough, Scott Schnell, Morten Schlütter, Philip A. Lutgendorf, Gary A. Tubb, Robert A.F. Thurman, and Indira V. Peterson. Last but not the least, my sincere thanks to the librarians at the University of Iowa (Edward Miner), Rutgers University (Triveni

Kuchi), and North Carolina State University (Darby Orcutt, Debbie Currie, and Avinash Maheshwary) who ensured that materials in Indian languages were available for my research.

The book is dedicated to my wife Sonia and our two boys Atharv and Anav, my little world that constantly reminds me of my dharma for them and for our beloved planet for all our future generations.

October 2, 2010
(The Birthday of Mahatma Gandhi)
Texas, USA

List of Abbreviations

AV	Atharv Veda
BhG	Bhagvad Gītā
BhP	Bhāgavata-Purāṇa
MBh	Mahābhārata
MSm	Manu Smṛti
ṚV	Ṛg Veda
ŚBr	Śatapatha Brāhmaṇa
SV	Sāma Veda
YV	Yajur Veda

Chapter 1
Introduction

Dharma is ethics and ethics dharma.

Ninian Smart[1] (1993)

This is a book about the lives of villagers in the states of Rajasthan and Gujarat in India. Having lived in India in my adolescence, I had sometimes wondered about the relationship of Indian villagers with their natural resources. My interest in this study was sparked in spring 2005 when I took a course on religion and environmental ethics with Professors Frederick Smith and Scott Schnell at the University of Iowa. This began my interest in exploring the diverse ways different communities interact with their environment. Out of thousands of different castes, groups, and sects present in the Indian villages, I have picked three communities spread across the villages of Rajasthan and Gujarat that have not only worshipped nature but also tried to protect and conserve it based on their beliefs and practices.

In Indic traditions, we see a number of rituals and myths in which mountains, rivers, trees, animals, and birds are revered. Despite this nature worship in India, its natural resources are under heavy pressure with its growing economy and exploding population. This has led several scholars to raise questions about the role religious communities can play in environmentalism. Does nature worship inspire Hindus to act in an environmentally conscious way? Is there any relationship between their reverence for bio-divinity[2] and their care for biodiversity? Since the 1990s, India has been embracing capitalism, consumerism, and urban development at a rapid pace. At this turning point, will it join the other developed countries as a major contributor to global warming?

In this book, I have tried to explore some of these questions with three communities, the Swadhyaya movement, the Bishnoi community, and the Bhil tribe. While the first one, the Swadhyaya movement, arose in the mid-twentieth century in Gujarat as a "New Religious Movement,"[3] the origin of the latter two dates back to medieval times in Rajasthan as two distinct "tribes" or "castes." None of these groups arose as a reaction to "global warming" or "saving biodiversity." Such a reaction has always been the motivation for environmentalists to launch

---

1 Cited by Bilimoria et al. (2007)

2 I borrow this term from Emma Tomalin (2009). I review her observations later.

3 "New Religious Movement" is a sociological term used to describe newer movements within established religious traditions. I describe and question this category later.

their campaigns in these times of "Inconvenient Truth" and yet some examples of ecological activism set by these three communities seem more effective than many governmental initiatives. The Swadhyayis (followers and practitioners of the Swadhyaya movement) have built their tree-temples and water-harvesting sites in Gujarat and other states with a high success rate. Bishnois have been active protectors of their natural resources. Similarly, Bhils have protected their sacred groves even when people destroyed the surrounding forests in Southern Rajasthan (Haynes 1998).

I present each of my case studies in separate chapters beginning with the Swadhyaya movement, followed by the Bishnoi community, and then the Bhils. Following David Haberman and Lance Nelson, I argue that "anyone wishing to understand the relation between religion and ecology in India, or to think or act ecologically in an authentically Hindu context, must come to grips with the mythic and sacred dimensions within which Hindus function – and the ecological implications thereof" (Haberman 2006: 2). Thus, before I initiate my discussion on the environmental practices of these communities, I explore the complex religious world of each, exploring their history, theology, and rituals that I encountered while reading their texts and talking with their followers, priests, and other volunteers during my visits to various temples and pilgrimage sites in their villages. I also question some of the categories such as "religion" and "New Religious Movement" as applied to these communities throughout my descriptions.

After Robert Redfield and M.N. Srinivas, Milton Singer (1972) explored the dichotomy of "great" and "little" traditions of India differentiating the "Brahmanical Hinduism" from "Local Hinduism." Following Singer, I apply this dichotomy of great and little traditions to modern and traditional "Hinduisms."[4] Modern "Hinduism" includes the movements such as ISKCON, Ramakrishna Mission, Transcendental Meditation, and Art of Living among others.[5] On the other hand, traditional rural "Hindu Dharma"[6] includes the rural communities such as Swadhyayis, Bishnois, and Bhils, that are largely present in villages and

---

[4] This dichotomy has also been redeployed by ethnomusicologist Peter Manuel to capture the influence of mass-mediated popular culture in contemporary South Asia. See his *Cassette Culture: Popular Music and Technology in North India* (1993).

[5] Following Axel Michaels, I am categorizing these groups under the category "modern" based on the fact that they were all founded in India but successfully spread in Western countries based either on their English literature often written by their founding gurus or their Western-style professional management including PR campaigns and websites (Michaels and Harshav 2004: 22).

[6] I differentiate "Hinduism" as a "Religion" from "Hindu Dharma" later based on the distinction between urban Indians' modernized and organized religious communities from the rural Indians' traditional "way of life." This distinction is also based on the use of languages in these two categories. The urban religious organizations largely use English while the rural "dharmic" communities continue to use their vernacular languages.

small towns of India. While my research with rural communities had little, if any, evidence of their ecological practices being influenced by modern scientific researches about global warming, the modern organizations, on the other hand, seem to be largely responding in their own ways to "save the planet," joining the global awareness movement. In this way, I present Indian communities in two different models: "great" and "little." The former is modern, English-speaking, urban-based, and fully conscious about environmentalism in its list of social causes and the latter is traditional, vernacular, rural-based, and only somewhat conscious about environmentalism.

These two kinds of environmentalisms then raise a bigger fundamental question. What is the difference between the two? In my chapters about each community, I show that there is no category of "environmentalism" in the "way of life" of traditional Indians living in the villages. Instead of using the categories of "environmental ethics" and "religion" to interpret their "way of life," I suggest that "dharma" presents a better alternative. Here I have tried to build on the seminal work of two scholars, Ariel Glucklich (1994) and Austin Creel (1977). Glucklich saw dharma as a phenomenon that helps Indians transcend the boundary of subject and object. I suggest that this holistic attitude of Indians based on dharma can be used for wider environmental awareness as has been done to some extent in the Swadhyaya movement. Creel, on the other hand, tried to connect the Western notions of ethics with the Indic notions of dharma. He observed that while several modern Indian thinkers have written about dharma, they are yet to develop a comprehensive ethical framework based on dharma. He suggested that this could be done by wide-ranging reinterpretations of dharma in line with modern developments. Following his suggestions, I try to correlate the discourses of the founders of the communities of Bishnois and Swadhyaya with ethics, especially as it relates to natural resources. I show that both these gurus succeeded in mobilizing and ethicizing their communities based on their dharmic teachings. In this way, building on Creel and Glucklich, I argue that dharma could be developed as an alternative sociological and anthropological category to study Indic traditions. Following Weightman and Pandey (1978), I argue that the concept of dharma can be successfully applied as an overarching term for the sustainability of the ecology, environmental ethics, and the religious lives of Indian villagers. The distinct categories of "religion," "ethics," and "ecology" work well for the "modern" urban Indians. However, for millions of rural Indians, "dharma" unites and synthesizes their way of life with environmental ethics, as also noted by Ninian Smart in his quotation above. Thus, I have tried to follow the project of McKim Marriott (1990) who used the term *ethnosociology*, the alternative disciplines of social science, instead of juxtaposing the Western categories on the non-Western societies.

Thus, my study of different examples of environmentalisms in the Indian villages raises several questions. Is it possible to transcend this dichotomy of tradition and modernity? Can we modernize and rationalize the traditional communities or can we "traditionalize" the modern communities? Which approach is better for the future? I present my conclusions in light of these questions. My hypothesis is that for the Indian traditional communities, environmentalism is ingrained in their daily "way of life" and their religious ethos that they often describe in vernacular languages using the term dharma that has multiple meanings including ethics, virtue, religion, sustenance, order, and law (Edgerton 1942). Based on this overarching presence of "dharma," I suggest that Dharmic Ecology of these communities offers a unique avenue for approaching environmental restoration today. Before I get to my three case studies in next three chapters, first, I review the existing scholarly literature about Indic Traditions and Environmental Ethics.

Chapter 2
Theoretical Propositions for Indic Traditions and Ecology

Environmentalism is comparable to a child that only recently learned to walk. Ecospiritualities of different kinds seem to be the invisible backbone of the growth of this child.

Bergmann (2006)

Lynn White is widely considered the pioneering scholar who criticized Western religions for the current ecological problems in the world (1967). The summary of his argument is as follows. Humans have tampered with nature unlike any other species, e.g., the Romans cut the forests to build their ships and the British built the Aswan Dam. Similarly, deforestation and erosion resulted from mining for potash, sulfur, iron ore and charcoal. Population explosion, sewage deposits, and garbage deposits are other major problems that continue to impact global ecology catastrophically. According to White, descendents of Northern European peasants turned out to be the biggest exploiters of nature with their scientific and technological progress in modern times. The transformation of humans from being part of nature to exploiter also coincided with another major shift in the human psyche in Europe. This was also the period when Christianity was "destroying" pagan animism there. According to White's interpretation of the Bible, God made humans in his image and they are supposed to dominate nature for their proper ends. By transferring all the divinity from nature to a superhuman God in heaven, "Christianity made it possible to exploit nature in a mood of indifference to the feelings of natural objects" (1967: 1205). Let me now analyze some of White's conclusions.

Christianity located the divinity in the God as a trinity – Father, Son, and Holy Spirit. The Bible does say that God made humans in his image but to interpret it to mean that God meant humans to dominate nature is an over-simplification of the Bible's story of Genesis. My point is that the Bible might have rendered the nature as profane but it did not exhort humans to exploit it for their selfish greed. White goes to the extreme of attacking the Bible in his agenda of searching for the roots of the current ecological crisis and misinterprets the story of Genesis. White falls in the category of those Orientalists who in their romantic image of Eastern religions saw only mistakes in the Western religions. J.J. Clarke (1997), George A. James (1999) and others have noted the two extremes. On the one hand, they seek all the solutions in Eastern religions and see only problems in Western religions. On the other hand, some regard non-Western traditions to be too irrational and superstitious and portray Western traditions to be more "scientific."

Overall, White tries to build a good case against the Judeo-Christian religious tradition and its aftermath, leading to the current ecological crisis. As I noted above, he leaves some holes in his thesis. I agree with White's conclusion that there is a need to search for fine examples within each religious tradition such as Saint Francis of Assisi in Europe and Zen Buddhism in Japan. Unfortunately, both the Western religious community and the scientific community seem to have ignored such examples at least until the late twentieth century.[1] For instance, the Jewish text Kabbalah employs the image of a cosmic chain in which every being is interlinked. All the elements of existence – from the most hidden to the most visible – are intimately bound to one another. All things trace their roots back to the inner recesses of the source of all being, Ein-Sof. Similarly, according to the Midrash's interpretation of the Bible, the first act Israel performs after entering the holy land is planting trees in their attempt to "walk after" God. God created the Garden of Eden so that men in his image can plant trees (Holtz 1984: 201–326). These are the examples that can build bridges among environmentalists and religious communities and go against the dominant interpretation of Genesis as presented by White.

I agree with White that each tradition is uniquely linked to its environment and it is best to try to look into native tradition for the ways each tradition or lineage absorbs the environment around it. Building on White's (and other scholars') recommendations, I now look into the case of Indic traditions and their potential environmental ethics. Scholars of environmental ethics and Indic traditions have differentiated two models of environmental awareness for India: "devotional model" and "renouncer model." These two models are based on a long-standing "snake-mongoose conflict" of India between the householders and ascetics that I have described elsewhere (Jain 2006). Householders perform devotional and ritualistic activities whereas ascetics perform austere practices.

Scholars such as Anne Feldhaus, David Haberman, and Vasudha Narayanan put the devotional model forward. Feldhaus notes that the Indologists have ignored the importance of worldly values in Indic traditions in general and the role of rivers for these values in particular (1995). She notes that rivers have been associated with the householder values such as wealth, beauty, long life, good health, food, love, and the birth of the children, rather than the ascetic values. Narayanan also distinguishes the ascetic model from the devotional model (1997). Narayanan argues that the ascetic model was largely limited to intellectual elites and their philosophical discussions. According to Narayanan, the Indic philosophical texts were never popular choices for oral and public religious performances. Hence, the world-denying attributes of Vedanta, as suggested by Nelson (1998) and described in philosophical texts of Śaṅkara, have only limited influence on the population outside the elite classes. What did reach the general population for thousands of years were texts such as the Rāmāyaṇa,

[1] As I describe later, several theologians and religion scholars have reinterpreted their traditions in their efforts "to become green."

the MBh, and the Purāṇas that were sung by bards, performed by dancers, and narrated by local storytellers.

Narayanan's argument in general is that even though the ascetic model advocates non-dualism, dharma texts maintain all the hierarchies of Indian society, namely those of gender, caste, and family. While ascetics may reject family life completely, dharma texts encourage one to earn wealth and procreate. Thus, her argument is that dharma texts are much more world-affirming and hence can be used to generate ecological awareness. Haberman also makes similar points (1994). Mumme notes a similar trend in Śrivaiśṇava Theology founded by Rāmānuja (1998). Further, Narayanan cites several passages from the Rāmāyaṇa and the Purāṇas that have statements such as "one tree is equal to ten sons." Several of these dharma texts have revered plants, animals, and birds.

Narayanan goes on to describe the edicts of Aśoka and the Artha Śāstra of Chāṇakya that describe several actions taken by the king or suggested for kings. The Artha Śāstra describes severe punishments and penalties for anyone cutting a branch or a tree. She then describes several examples based on Purāṇic deities' role in ecological awareness in South India, the biggest example being the Tirupati temple in Andhra Pradesh. This is generally called the richest temple of India, boasting of donations running into several hundred million rupees every year. What is remarkable is that this temple has changed its usual prasād of sweets into small plants that are given to devotees to be planted. This has resulted in several millions of plants planted to stop deforestation in the hills near Tirupati.[2] A similar initiative is taken by the G.B. Pant Institute of Himalayan Ecology that has been working with the temple of Badrinath. Its scientists produce the saplings; the priests bless them and distribute them as prasād to pilgrims.

Narayanan underscores that such ritualistic devotion toward a deity is a Purāṇic concept, different from Upaniṣadic metaphysical philosophies. She also notes that the Purāṇas and the epics describe the Ganga and other rivers of India as goddesses. In addition, she describes the "pregnancy-cravings" of the rivers and the offerings by devotees for such "cravings" in specific seasons.[3] Similarly,

[2] See the temple website for details: www.tirumala.org/activities_social_haritha.htm. Narayana has written about this in her Washington Post blog (viewed on April 3, 2007), http://newsweek.washingtonpost.com/onfaith/vasudha_narayanan/2007/02/one_tree_is_equal_to.html. However, recent visitors to this temple have not noticed this effort (personal communication with Frederick Smith, January 2008).

[3] The Dharma Śāstras, MSm, for instance, describes a highly symbolic worldview consisting of milk-bleeding trees that mark boundaries, wives who are identified with fields and hunted deer, water tanks that must be purified like persons, rivers that menstruate, and others (8.44, 5.108). Consider, for instance, the following rule in Kātyāyana Smṛti about river bathing: "In two months, beginning with Śrāvaṇa, all the rivers get their menstrual courses. No one shall bathe in them, excluding the rivers which go to an ocean (10.5)" (Glucklich 1994: 13). Anne Feldhaus (1995) has also described similar patterns in the rivers of Maharashtra.

several temples in Kerala dedicated to Purāṇic deities have a Sthala-Vṛkṣa within their compounds, thus tree and temple are inseparable entities. She also cites some examples of how traditional dances portray stories and songs to spread ecological awareness, including dances by the well-known Indian dancer Mallika Sarabhai. All of these activities are examples of the devotional model leading to ecological awareness and reverence. However, Narayanan echoes Kelly Alley and Vijaya Nagarajan, who have observed the ill effects of the devotional model. She cautions that the devotional model sometimes transgresses the boundaries of pure versus auspicious and impure versus inauspicious. Kelly Alley has observed in Banaras that Hindu devotees note that the Ganga is polluted but for them Ganga being a purifying goddess can never be inauspicious (2002). The goddess Ganga is supposed to clean both the river and the devotees. Thus, the devotional model can sometimes turn into a complacent model. Similarly, Nagarajan has noted this for the goddess earth in South Indian villages (1998).

Turning to the ascetic model, Chapple, principally a scholar of Jain tradition, has advocated that non-violence to animals, trees, and self, combined with non-possessiveness can result in ecological awareness (1993: 73):

> [T]he solutions that Gandhi proposed to counter the ills of colonialism can also be put into effect to redress this new and ultimately deleterious situation. The observance of nonviolence, coupled with a commitment to minimize consumption of natural resources, can contribute to restoring and maintaining an ecological balance.

Chapple notes Gandhi as an example who limited his possessions and "vital needs" and thus can serve as an inspiration for ecological ethics. According to Chapple, Gandhi and others who follow the ascetic and yogic values such as truth, non-stealing, non-possession, celibacy, and non-violence, serve as role models for limiting the consumption and thus reducing the burden on ecology (1993: 71). Vinay Lal and others have also put forward Gandhi as "too deep" even for "deep ecology" (2000). What Lal means is that Gandhi serves as a role model of practicing an environmentalism that is much beyond what "deep ecology" presents in its philosophy.

Chapple and scholars of ascetic traditions have recommended the ascetic model (2002). Chapple notes that Jains, following their ascetic values, have exerted an active social conscience (Martin and Runzo 2001):

> They successfully convinced the first Buddhist monks to cease their wanderings during the rainy season, to avoid harm to the many insects and plants that sprout during the monsoons. The Jaina community has developed and implemented lay codes for assuring an integration of nonviolent values into the workplace. Jainas have lobbied against nuclear weaponry. The head of the Terapanthi Shvetambara sect, Acharya Tulsi, took a public stance on numerous issues.

Against Lance Nelson's questioning of world-negating Vedanta and the renouncer model (1998), Eliot Deutsch observes (1970: 4): "Vedanta would maintain that fundamentally all life is one, that in essence everything is reality; that Brahman, the oneness of reality, is the most fundamental ground of all existence. Being free, the self of the individual can behave as if unattached and without destructive intentions." Therefore, Deutsch concludes, "paradoxically, when nature is seen to be valueless in the most radical way, it can be made valuable with us in creative play" (1989: 264). Chapple's observations about the Jain community seem to match with Deutsch's argument that the practitioners of the renouncer traditions can also be proactive about the ecological concerns, although this is questioned by Nelson.

I agree that following the ascetic model may result in decreasing one's "vital" needs and thus limit consumption and consumerism. However, in a country such as India, the attention and preference of society today is not in limiting consumerism. Its burgeoning middle class and vibrant cities are embracing consumerism at the fastest pace in their history. To advocate that the lower and middle classes accept ascetic lifestyles seems anachronistic at this stage. On the other hand, the devotional model seems to apply more easily to them as seen by the success of the Swadhyaya movement in Gujarat, Maharashtra, and other parts of India, as I describe in a later chapter. Nitin Sethi and Ramya Viswanath had similar observations about the loss of sacred groves in India in their article in an Indian environmental magazine *Down to Earth* (2003):

> The force today is the economy; monetization is transforming Indian villages like never before. It changes the atmosphere in which the society constructs its ways of self-control and codes of morality – the myths, legends, gods and icons. What we need to make sure is that social norms that accompany the new economy, again accept groves and forests as central to the lifeworld. Ecological sense must fit with economic prudence. And in turn, we should reset the price for deviating from ecological prudence.

Bishnoi Guru Jambheśvara, in fifteenth-century Rajasthan, was probably the first Indian guru to emphasize ecological awareness in his teachings. In his 29 rules, he specifically prohibits harming trees or animals, and encourages vegetarianism. Of the 29 rules laid down by Guru Jambheśvara, eight rules have been prescribed to preserve biodiversity, and encourage good animal husbandry. These include non-sterilization of bulls and keeping the male goats in a sanctuary, prohibition against killing the animals and the cutting down of any type of green trees and providing protection to all life forms. The followers are even directed to see that the firewood is devoid of small insects before burning it in their hearths. Even the wearing of blue cloths is prohibited, because the dyes for coloring them used to be obtained by cutting a large quantity of shrubs. In the eighteenth century, 363 Bishnoi people sacrificed their lives to protect a desert tree, khejari, from the soldiers of the Jodhpur king. Today, the Indian government has a special ecological award in honor of Amrita Devi, a Bishnoi woman who can be referred to as the first ecological martyr of India. I describe the Bishnoi community in detail in a later chapter. The Chipko movement, led by Sunderlal Bahuguna and Chandiprasad Bhatt, was allegedly inspired by a Bishnoi sacrifice that in turn inspired the Appiko movement of Karnataka.[4] In both these movements, villagers embraced the trees to save them from the forest department's assault for timber. Once again, in these examples we see devotional texts such as the BhG and the Purāṇas being used and recited during their ecological activism. Ladies tied Rākhi on trees in Uttaranchal and sung verses from the BhP for several days when Bahuguna was fasting in one of the key events of their movement, as explained by George James (2000).

The above examples suggest that there is a need to combine both the models of devotion and asceticism. Gerald Larson cautioned against making any boundaries among disciplines in the study of ecological problems (1989). I would like to extend his suggestion to create a symbiotic model using both asceticism and devotion. As suggested by Hawley (1988), many devotional sects, such as Bishnoi, were actually founded or spread by ascetics. Gurus and leaders such as Jambheśvara, Gandhi, Bahuguna, and Athavale have contributed to India's ecological awareness in a significant way, so we need to learn and combine these models. T.N. Madan had cited J.L. Austin (1987), "The history of Western philosophy was littered with 'tidy-looking dichotomies', with the student being required to embrace one half or the other." Following Austin, I would like to transcend and merge the dichotomous categories of devotional and ascetic models. Two Indian anthropologists, L.P. Vidyarthi and R.S. Mann, had developed a theoretical model called the Nature-Man-Spirit complex to study tribal societies and their dependence on rituals and vice versa. I think this model may prove to be a useful one. In a volume edited by R.S. Mann (1981), different scholars wrote about tribal communities of India in the Andaman and Nicobar Islands, Uttaranchal, Rajasthan, Bihar, and elsewhere. The connections among humans, ecology, and traditions in these societies are a manifestation of their "way of life." For them, there is no category of "religion" or "environmentalism." This resembles the Swadhyayis, the Bishnois, and the Bhils who also are barely conscious about their environmentalism, as I will show in my case studies.

David Haberman differentiates two theological notions of environmentalism: transcendent and immanent (2006). Haberman summarizes the colonial agenda to reduce Indic traditions to a transcendental and otherworldly "religion." He cites Ronald Inden to show the political agenda behind this: "this was important for the imperial project of the British as it appeared, piecemeal, in the course of the nineteenth century. Because the theist creeds and sects, activist and realist, were the world-ordering religions of precisely those in the Indian populace,

[4] See www.youtube.com/watch?v=3zuqvGAzDi0 for a documentary about the Appiko movement and its inspiration from the Chipko movement led by Sunderlal Bahuguna.

among the Hindus, that the British themselves were in the process of displacing the rulers of India" (Inden 2001). Thus, we find a scholarly trend that sees only ascetic and world-negating examples from diverse and pluralistic Indic traditions. This project to highlight the world-negating philosophy of Śaṅkara ignored most of the theistic examples from the Purāṇas. Haberman cites several examples of the colonial period to show how colonial scholars chose Śaṅkara's transcendental philosophy as their preferred icon of the Indic traditions. Thus, according to Ronald Inden, Śaṅkara became their "hero of imagination." According to Haberman, this did injustice to other immanent philosophies and theologies of Indian traditions, such as the Puṣṭī Mārga of Vallabha and Vaiśnavism of Rāmānuja (Smith 2005). Haberman also convincingly shows how the residual effects of this colonial agenda percolate in the current writings of Lance Nelson, Warwick Fox, and proponents of "deep ecology," including Tom Hayden and Arne Naess. According to Haberman, even in this era of post-colonial scholarship, several scholars have continued to follow the agenda set by Orientalist scholars such as William Jones, Henry Colebrooke, Paul Deussen, and Max Muller. This reductionistic interpretation of Indic traditions leads some observers to assert that "Hinduism," being a world-negating religion, cannot inspire environmentalism. Like several Orientalists, Naess also reduced "Hinduism" merely to a philosophical tradition of advaita and self-realization, ignoring several diverse traditions with theistic and ritualistic elements. Haberman critiqued this trend (2006) and demonstrated that most Hindus identify themselves with the theistic, Purāṇic, and world-affirming traditions. As a counterpart to the theoretical framework established earlier, Haberman concludes his book with numerous examples of Indian environmentalists. Their major inspiration comes from Hindu theistic traditions, such as the Chipko movement of Uttaranchal, Sunderlal Bahuguna's struggle against the Tehri Dam, Friends of Vrindaban, Veerbhadra Mishra's Sankat Mochan Foundation, and Eco-friends in Kanpur, just to name a few from his many examples.

To his list, I can add the Athavale's Swadhyaya work that relates to the environment in its own unique ways. Interestingly, in several of his discourses Athavale paid glowing tributes to Śaṅkara. According to Athavale, Śaṅkara was not just an ascetic world-denying philosopher, as is described in academic writings. For Athavale, Śaṅkara and several other Indian gurus and sages were world-affirming people involved with uplifting and reforming the Indian society of their times. Athavale cited the Śaṅkara-Digvijaya as a historic recording of Śaṅkara's life that describes his debates and discussions with several thinkers of different ideologies and philosophies. According to this narrative, Śaṅkara "defeats" them all one by one in order to restore the Vedic tradition in India.

In this way, Śaṅkara's national travel to engage in philosophical debates with rival philosophers and thinkers is generally believed to have played a key role in re-establishing Vedic culture in post-Buddhist India. Indians imagine Śaṅkara as an ascetic but they also consider him one of the greatest Hindu revivalists. The legend of Śaṅkara is best captured in various versions of the Śaṅkara-Digvijaya, literally, Śaṅkara the universal conqueror (Pande 1994). Here, Śaṅkara is a campaigner for Hindu tradition whose mission was to re-establish Hindu tradition as the dominant tradition and remove Buddhist ideas from India. To that end, he traveled throughout India on foot and established four different Hindu Mathas in the Northern, Eastern, Southern, and Western ends of India. He defeated various Buddhist scholars in intense debates and converted them back to Hindu tradition. Stories of Śaṅkara's victory over Buddhist tradition are popular even in Nepal (Greenwold 1974).[5] The most interesting aspect of the Śaṅkara folklore of Nepal is that according to these stories, Śaṅkara forces the Buddhist monks to become householders who continue to serve as temple priests. Śaṅkara does not encourage renunciation or world-denial in Nepal. It is this popular version of Śaṅkara that is completely ignored by scholars. If the world was an illusion, māyā, for Śaṅkara why would he work to "defeat" Buddhist tradition and other ideologies in the popular discourses as captured in the Śaṅkara-Digvijaya. Moreover, numerous devotional hymns, such as the Bhaja Govindam, are also attributed to Śaṅkara by lay Hindus. Once again, this shows that Śaṅkara is imagined as a theistic Hindu in popular imagination, not just a world-renouncing ascetic philosopher as scholars describe him. This theistic model is successfully employed in environmentalism inspired by dharmic traditions, by several activists and leaders as listed by Haberman as well as in the Swadhyaya movement and in the Bishnoi community.

Contrary to religious environmentalism inspired by Indic traditions as described by Haberman, Chapple, and Narayanan, Emma Tomalin is skeptical of the role of religion in environmentalism (2004). I now summarize her paper and then present my responses afterwards. In about 30 pages, she touches several issues. First, she differentiates "recognition of bio-divinity" in several traditions from "religious environmentalism." She argues that since the latter is a product of post-materialist environmentalist philosophy emerging from the West after the 1960s, the former cannot be assumed to respond to the global ecological problems, given its lack of awareness and adaptability to the modern problems. She agrees that the recognition of bio-divinity easily finds support from within "the Hindu tradition," but there is an immense difference between the priorities and concerns of the modern environmentalist and the worldviews of much earlier Hindu thinkers. She also questions whether Hindus in modern India, whom she calls "environmentally illiterate" with little or no knowledge of the language and concepts central to contemporary environmentalist thinking, actually share the religious environmentalist's goal of ecological sustainability. She also argues that the "subaltern" poor of India cannot afford to put the "Earth first" since they depend on nature for their daily survival. This is different from the Western "deep ecology," which aims to protect the earth from all human interventions. Establishing this East–West dichotomy, she contradicts herself and criticizes those authors who blame Western capitalism for global ecological

[5] I thank Michael Baltutis (University of Wisconsin at Oshkosh) for this reference.

problems. For her, the environmentalist movement is definitely a Western approach, but to blame the West for ecological problems is a "characteristic of post-colonial critique" which need not be taken seriously. Apparently, if she talks about a solution, she refers to it as "Western," but if she describes a problem, she argues against calling it "Western." Her approach of calling "subaltern" Hindus environmentally illiterate also exposes this bias. Depending on the per capita ecological usage of an average Indian, or any non-Western person for that matter, compared to the same usage by a person in the Western countries would make a Western lifestyle "environmentally illiterate," as I will show later following Ramachandra Guha.

She then goes on to note that "religious environmentalism offers an interpretation of tradition and not a traditional interpretation" (emphasis in original). As I will show in my chapters about the Swadhyayis and the Bishnois, their founder gurus inspired their followers to be protective and reverential toward ecology, based strictly on the traditional interpretation, i.e., just by recognizing the "bio-divinity" (as Tomalin would call it) and to practice it in their daily lives with greater awareness toward their environment. Unlike the late twentieth-century Christian thinkers, Indian gurus did not have to reinterpret their scriptures or traditions to make them more eco-friendly as was done by several Christian theologians in their responses to the criticism posed by Lynn White. Thus, Tomalin equates Christian interpretation of tradition with Indic traditions. However, contrary to her generalization, the traditional interpretation has inspired Bishnois, Swadhyayis, and others to become religious environmentalists, as I intend to show later. This will also refute her charge of calling such an exercise limited to only the "full-stomach" societies of the West and not the "empty-belly" societies of Asia.

She then chooses sacred groves as a key area where religion has entered the environmental debate in India by reviewing related scholarly writings. She notes, "To this day one can find patches of forest all over India that have been protected due to religious custom." These are described as "hotspots of biodiversity" and are claimed to be of ecological significance. She then goes on to describe the continuous depletion of these groves due to "the dilution of traditional values, Westernization, and migration." She also discovers with respect to Coorg sacred groves in Karnataka that "commercialization of the land has led to a transformation in 'traditional' farming practices and has brought changes to the institution of the sacred grove. Before the introduction of coffee into the region by the British in 1854, Coorg was almost completely covered with forest and in its Eastern region with thick jungle." This observation contradicts her earlier statements. Here, she seems to agree that the "traditional" values had preserved the groves while "Western" style commercialization and forest control has led to its destruction. Ironically, her own observations have failed to influence her conclusions throughout the article.

She then reviews the work of Freeman about the sacred groves of Kerala whose research suggests that sacred groves have always taken a variety of forms that do not necessarily coincide with the modern environmentalist's idea of "pristine relics from a primeval past." Any protection of biodiversity was coincidental rather than intentional and sacred groves were protected out of respect for the deity rather than due to an innate belief in the intrinsic value of nature. As I will show in my research about Swadhyayis (as builders of several new sacred groves and water harvesting sites in the twentieth century) and Bishnois and Bhils (as protectors of their older sacred groves), I largely agree with Freeman's observations. Juxtaposing modern ecological discourses onto traditional communities is somewhat unfair. For the traditional communities, their daily "way of life" was the focus, not the ecological discourse. It is only by the dharmic practices rooted in their traditions that they can help "save the biodiversity." Thus, their environmentalism will always remain a by-product of their dharma. She also cites JNU Professor P.S. Ramakrishnan in this regard. Instead of appreciating this traditional interrelationship of humans with ecology, Tomalin notes that the modern ecologists are more interested in taking a secular scientific approach to preserve the sacred groves in particular and ecology in general. This, in her view, limits the role of religious environmentalism. Again, she has missed the point that global ecology cannot be saved by armchair researchers alone. Ecology is an issue about the people, of the people, and for the people. If the scientific community wants to reach out to the people, they will have to appreciate the role indigenous traditions have played in maintaining the balance in much of the world, where Western style progress has not yet invaded. This is increasingly acknowledged by the scientific community, as is evident from the citations provided by Tomalin, but again this eludes from her conclusions, which either categorically reject the role of traditional values or equate it with the "Hindu Right," which is her next topic. She describes how the Hindu nationalist movement has shown interest in environmental issues such as the "Vrindavan Forest Revival Project," River Ganges, and the Tehri Dam. Here, I agree with her and can add the Rāma Setu issue also, in which the Hindu nationalists have utilized the environmentalist cause for their political agenda.[6] After ignoring observations from several of her respondents in her fieldwork, she concludes her article with strong assertions without any evidence to support them. She resorts to the outdated practice of looking for "great" and "little" traditions within India by separating indigenous traditions from Brahminical. Several scholars, including Frederick Smith, have noted that the so-called "great" and "little" Indian traditions cannot be studied as two dichotomous groups of traditions. Further, she has also unnecessarily brought in the issue of the "Hindu Right" as an obstacle for using traditional religious environmentalism. By this logic, religious environmentalism should be stopped across the world immediately because various "Right-Wing" organizations present in Christianity, Islam, and other religions would hijack ecological discourses for their own political agendas. She wants to replace

6 www.ramsethu.org (viewed on December 20, 2007).

the notion of "religious environmentalism" with "secular environmentalism." Against her concern about mixing religion with environmentalism, I found the approach of Douglas Johnston (2003) more relevant in which he urges for a long-term strategy of cultural engagement, backed by a deeper understanding of how others view the world and what is important to them. According to Johnston, in the non-Western cultures, religion is a primary motivation for political actions. Historically dismissed by Western policymakers as a divisive influence, religion in fact has significant potential for overcoming the obstacles that lead to paralysis and stalemate. The incorporation of religion as part of the solution to such problems is as simple as it is profound. It is long overdue. Bergmann, in his quote in the beginning of this chapter, also underscores spiritual traditions as the backbone for new emerging environmental movements across the world.

I tend to agree with Johnston that the older approach of applying the Western model to try to solve purely from a secular framework has resulted in more problems than solutions. For most of the non-Western rural societies, religious ethos is an integral part of their "way of life." As I show later, in my case studies of Bishnois and Swadhyayis, I found that their environmental initiatives were inspired by the traditional teaching and guidance of their founding gurus rooted in religious ethos.

Overall, although I disagree with Tomalin on some issues, I reviewed her work here because it provides a good overview of different problems with the category of religion and environmentalism in the Indian context that is my focus in this book. I also agree with her arguments about "bio-divinity" which does not automatically lead to environmentalism for preserving "biodiversity." In my own fieldwork with the Swadhyayis and the Bishnois, I often found that environmentalism is merely a by-product of their traditional lifestyles revering the bio-divinity. For them, biodiversity is merely a by-product of their means and goals of practicing their dharma.

I would like to deal with the issue of connecting "religion" with "ecology" in all my case studies. Specifically, my argument would be that it is a modern concept for Indian organizations to see religion and ecology as two separate and distinct notions. The case studies that I describe do not see religion and ecology as separate. For them, their "way of life" implicitly includes environmentalism. They revere nature not for saving biodiversity but because this is how they practice their traditions and their dharma, or dharam, as they describe it in vernacular Indian languages. As mentioned earlier, the Nature-Man-Spirit complex model developed by Vidyarthi and Mann also notes the intertwined nature of spiritual practices of several tribes that includes environmentalism instead of seeing it as a distinct philosophy. Traditional communities practice environmentalism by practicing their "Dharma" which in many ways is different from "Religion." The Swadhyaya and the Bishnoi communities were founded by their charismatic gurus who adopted several Indic traditions in their novel ways to preach and spread their new ideologies. In contrast to these two communities founded by their gurus, the case of the Bhils of Rajasthan presents yet another example of a traditional community practicing their rituals and indirectly relating to environmentalism as I show in a later chapter.

Chapter 3
The Swadhyaya Movement[1]

There is a divine power in trees that makes it possible for water and fertilizer to rise from the roots below and reach the top portion against the gravitational force. It is not just the result of Keśākarṣaṇa *(capillary action) but it is* Keśavākarṣaṇa *(Kṛṣṇa's force).* Vṛkṣa main Vāsudeva *(Kṛṣṇa in trees),* Paudhe main Prabhu *(Almighty in plants)!*

Having heard about the tree-temples of Swadhyaya, I called their office in Mumbai to visit one such site in summer of 2006 during my trip to India. Soon, I found myself on my way to Valsad in Gujarat. I arrived at the home of a local Swadhyaya volunteer, Maheshbhai, who was supposed to take me to the site. Several other people had also come to accompany me to the site. All of them showed warmth and enthusiasm to welcome me and explain to me about tree-temples and several other works and ideologies of the Swadhyaya movement. The tree-temple, established on June 22, 1987, appeared as if an oasis has suddenly sprung up out of nowhere. It was a dense garden of trees of mangoes and *chikoo* (sapodilla). I appreciated the view of lush green trees and was particularly impressed that it was built on a land where people had just lost all hopes of cultivation. Even the government had declared it as a barren land. As the caretakers of this garden began explaining about the way they perceive the trees and the vision of their guru Athavale with the words mentioned in the quotation above, I began asking questions related to environmentalism. What I present below is based on several such interviews with Swadhyaya followers who have participated in such work. I have also extracted relevant information from the vernacular literature of Swadhyaya that is based on the video-recorded discourses of Athavale.

Swadhyaya is one of the least known "New Religious Movements" that arose in the mid-twentieth century in the Western states of India. Although this movement now has some presence in several Western countries such as the United States, Canada, and the UK, it has not received the attention of scholars of Hinduism except in a few introductory articles.[2] Before I begin introducing the origin and history of Swadhyaya, it is important to mention that its projects are

---

1 Portions from this chapter have been published in an article in the journal *Worldviews: Global Religions, Culture, and Ecology*, Vol. 13, No. 3, 2009, pp. 305–20.

2 Between 1994 and 1996, some scholars had visited the Swadhyaya villages. Their observations were compiled in an edited volume by R.K. Srivastava (1998). This is a helpful introduction of the movement. In addition, I have provided some other scholarly articles in the bibliography, see Dharampal-Frick (2001), James (2005), Little (1995), Paranjape

not declared as "environmental." In fact, one of the Swadhyayis (practitioners of Swadhyaya) was taken aback when I told him about my topic of research. His apprehension was that I might misrepresent Swadhyaya if I choose to research it from an ecological perspective. According to him, Swadhyaya and its activities were only about their devotion to the Almighty; ecology is not their concern. Environmental problems are due to industrialization and the solution lies beyond Swadhyaya's activities. Swadhyayis are not environmentalists!

Athavale had repeatedly emphasized that the main goal of Swadhyaya is to transform the human society based on the Upanishadic concept of "Indwelling God." According to him, since the Almighty resides in everybody, one should develop a sense of *spiritual* self-respect[3] for oneself irrespective of *materialistic* prestige or possessions. In addition to one's own dignity, the concept of "Indwelling God" also helps transcend the divisions of class, caste, and religion and Athavale exhorted his followers to develop the Swadhyaya community based on the idea of "brotherhood of humans under the fatherhood of God." Activities of Swadhyaya are woven around this main principle, which in turn are also aimed at the Indian cultural renaissance.

Although environmentalism is neither the means nor the goal of Swadhyaya's activities, natural resources such as the earth, the water, the trees, and the cattle are revered and nurtured by Swadhyayis based on this understanding. Environmentalism does come out as an important by-product of its multi-faceted activities and this was noted by a 1992 conference in Montreal where Swadhyaya was invited to present its ecological philosophy and work.[4] I argue that a multivalent term like dharma can comprehend and describe this kaleidoscopic phenomenon and the way it relates to the ecology. Swadhyaya followers do not regard environmentalism as their main duty, their dharma. Alternatively, from the outside, one can regard their dharma, their cultural practices, as ecologically sustainable as I show below. I also want to note that my observations are based on their activities in the rural parts of India. The urban and diaspora Swadhyayis do not have such ecological projects.

(2005), Rukmani (1999), Sharma (1999), and Unterberger et al. (1990). Also, see Ananta Giri's monograph (2009).

[3] Since the 1990s, this became the theme of celebration for Athavale's birthday, which is celebrated as "Human Dignity Day" around the world.

[4] This conference, "LIVING WITH THE EARTH" – Cross-cultural perspectives on sustainable development: indigenous and alternative practices, took place on April 30–May 3, 1992. It was organized by the Intercultural Institute of Montreal, Canada. A three-page report titled "Presentation by Dīdī on the Swadhyaya Movement" was written by Robert Vachon in the proceedings of the conference. While in Montreal, David Cayley had interviewed Dīdī as well as Dr Majid Rahnema. This was subsequently broadcast nationwide by the CBC (Canadian Broadcasting Corp) on a one-hour radio program called "IDEAS."

From 1942 onwards, Athavale started giving discourses on the Vedas, the Upaniṣadas, and the BhG for several decades in Madhav Bagh, a section of a crowded area called Bhuleśvara, in south central Mumbai until his death in 2003. The Swadhyaya movement has multiple origins and beginnings. In one sense, it started when its founder started giving his discourses in 1942. Alternatively, from the socio-ecological perspective, Swadhyaya's work started in 1979.[5] Before describing the Swadhyaya work involving ecology, I provide an analysis of the ideology, history, and processes of Swadhyaya from the theoretical framework of "New Religious Movements (NRM)." This will also exemplify how Swadhyaya has developed its vision encompassing spiritual, social, economical, and ecological upliftment of Indian villages.

Although Swadhyaya derives its inspiration from Indian cultural heritage, it rejects the label "Hindu organization." It has maintained a distance from other "Hindu" organizations and prefers to focus on socio-spiritual grassroots work in thousands of villages across India (and Indian diaspora abroad). It started from a handful of villages in Gujarat and later evolved to Maharashtra, Andhra Pradesh, Madhya Pradesh, and other Indian states. Currently, its presence can be noted in about two dozen countries spread over Asia, North America, Europe, Africa, the Middle East, Australia, and the Caribbean. In the United States, it had a modest beginning in 1978 and now records its presence in more than 35 states as told to me by Rameshbhai, a senior Swadhyayi in the USA. One of the central themes of Athavale's discourses was to inspire Swadhyayis to be world-affirming and actively work to restore Vedic culture (Unterberger and Sharma 1990). To that end, Swadhyayis have taken several socio-economic projects based on devotion, such as farming, *Yogeśvara Kṛṣi* (named after Kṛṣṇa in the BhG) and fishery, *Matsyagandhā* (named after the mother of Vyāsa, the Vedic sage). Athavale called them *prayogs*, experiments conducted on human society. Their ecological impact, though not acknowledged by Swadhyaya, is important and noteworthy as I show below.

Several New Religious Movements (NRMs)[6] have attracted much interest in both the journalistic and academic community. NRMs have sprung up in almost every society in all periods of history even though the term is usually understood to denote modern movements. Much research has been published about NRMs in the United States, Europe, and Japan. However, recent Indian NRMs are still in need of serious scholarly attention. Many NRMs, such as Sarvodaya, BAPS, Art of Living, and Arsha Vidya have continued to escape scholarly attention. Before I present the ecological projects of Swadhyaya, it is important to describe in detail its ideological foundations. Swadhyaya presents many features that do not fit

---

[5] A Swadhyayi pointed out to me that Athavale took more than three decades to launch a socio-ecological activity, which shows that his focus was always to reform human society, not environmentalism.

[6] Scholars have used "New Religions" and "New Religious Movements (NRMs)" synonymously and I follow the same practice in this section.

into the broad category of "New Religious Movement" while its other features seem to be appropriate for this category. Thus, I discuss the category of NRM as it relates to Swadhyaya. Alternatively, I explore if it may be better to describe it as a dharmic community rather than a "religious movement." One of its unique aspects is its focus to connect the role of religion for secular ends including environmentalism. Thus, I suggest that the Swadhyaya's activities can be better categorized as dharmic activities rather than religious.

A Brief History of Swadhyaya

Like other NRMs, Swadhyaya (literally, self-study) was inspired by its founder, Pandurang Shastri Athavale, *Dādājī* (meaning "elder brother" in Marathi and "grandfather" in Hindi). He was born in 1920 in a Maharashtrian Brahmin family. His father had founded a traditional school called *Gītā Pāthshālā* (BhG School) to teach the BhG in 1926 in Mumbai. In 1942, when suddenly his father fell ill, young Athavale took over the responsibility and continued the school. Although a Maharashtrian by birth, he developed fluency in Gujarati since most of his listeners were Gujarati. Not much is known about the early years of Athavale's work that seems to be limited only to preaching based on traditional Indian texts. However, this was soon to be changed when he participated in a religion conference in Japan in 1954. This proved to be a turning point in his life. Apparently, he was challenged there to show any village in India where people live based on the BhG's message (Srivastava 1998: 198). After his return from Japan, he decided to develop a role model based on the traditional dharmic teachings. His mission to start the constructive work first took shape in 1956 when he launched *Tattvajñāna Vidyāpīṭha* residential school to study the Indic traditions and philosophies. This event can be considered one of the early milestones of his work.

For first few decades after starting his work, Athavale's focus was to teach and spread the doctrine of the "Indwelling God," the presence of divinity in human and other beings. For Athavale, *asmitā*,[7] self-esteem had always remained an important virtue to be developed based on the awareness of Indwelling God. Although the ancient Vedantic texts mention about Ātman within each living entity as being identical with Brahman – the all-pervading soul of the universe – such Brahminical concepts have largely been a subject of research and discussion among scholars rather than to be practiced in daily life by Indian masses living in villages. I should also note here that during this period, Indian villages were largely undeveloped and the fruits of Indian economic development had not yet reached them. Thus, Athavale must have faced an uphill task to infuse a sense of

[7] *Asmitā* is also mentioned in the Yogasūtra. In 1.17 it is mentioned as a positive attribute to the state of Samādhi and in 2.3 and 2.6, it is mentioned as a negative attribute to be cleansed from the mind.

self-esteem and faith on Indian cultural traditions in villagers struggling to make ends meet. When the Western societies were already in their "postmodern" phase of facing the environmental problems that arose due to industrialization, Indian villagers were still in their "premodern" phase of agricultural economy. What is noteworthy is that Athavale succeeded in his mission even in this scenario.

One of the most important activities used to expand the doctrine of Swadhyaya are visits by Swadhyayis to different neighborhoods. These have several versions in Swadhyaya. At the town or suburban level, they are called *Bhāva Pherī*, literally emotional visits, that last a couple of hours conducted by groups of two or three Swadhyayis. The bigger version, called *Bhakti Pherī*, literally devotional visits, last a couple of days, during which these groups are supposed to stay overnight with fellow member-families to conduct similar visits in distant towns. Before the annual convention and other festival celebrations, Swadhyayis also conduct *Tīrtha Yātrā*, pilgrimages. These are similar to Bhakti Pherī but usually are targeted toward inviting the neighboring people to a festival or an event organized by Swadhyaya. The diaspora Swadhyayis also conduct international Bhakti Pherī in which they go to neighboring countries. Currently, members from the USA go to Suriname, Fiji, Trinidad, Australia, and New Zealand and those from the UK go to the other European countries. All these different versions of devotional visits have several goals. First, they are the means to offer one's gratitude to God. Athavale used to quote from the BhG,[8] and interpret it to exhort his followers to the effect that those who spread Kṛṣṇa's words in wider communities are beloved devotees of Kṛṣṇa. Athavale compared this activity with that of Christian missionaries in Africa and Buddhist monastics in Asia. Second, they serve as a means to break the barriers of caste, class, or religion among Swadhyayis themselves. When they perform these activities for long durations together, they develop new egalitarian relationships. During these "devotional visits," they are supposed to stay with others as part of a larger spiritual community (*Sangha*) and participate in the daily liturgy of chanting, discourses, and visits to the people in the neighboring towns or villages. These visits are also identified by Athavale as the practical dimension of spiritual fasting. According to him, fasting (*upavāsa*) is not just meant to change one's dietary habits but literally means to approach God. According to him, the meaning of *upa* is "near" and the meaning of *vāsa* is "to stay." Hence, upavāsa means to stay close to God. Similarly, *Ekādaśī* means Hindus performing fasts on the eleventh day of every month according to the lunar Hindu calendar. Athavale reinterpreted the importance of the number 11 (Athavale 2000b):

---

[8] *Na ca tasmān manuṣyeṣu kaścin me priya-kṛttamah. Bhavitā na ca me tasmādanyāh priyataro bhuvi* (BhG 18.69) (Literally, one who preaches the message of the BhG for the benefit of all is most dear to Kṛṣṇa.) This verse was often cited by Athavale to inspire his followers to spread the message of BhG "unto the last."

> Eleven is the sum of our five sense organs, five motor-organs, and the mind with which we function. On the eleventh day of each fortnight, the eleven elements of our personality are to be offered to God. In this sense, fasting means non-feeding the senses, body and mind on worldly pleasures. In this way, fasting becomes a spiritual act.

From Athavale's practical experiences with the thousands of villagers that he met in his visits over several years, he first tried to infuse a sense of dignity in them. Although they lacked materialistic possessions, he made them aware of their spiritual "possession." This worked as a double-edge sword. It helped develop self-esteem and also helped spread Indian traditions in remote areas of Gujarat and Maharashtra. He used the concept of *Trikāla Sandhyā,* a threefold prayer mentioned in the Dharmaśāstras. While this is a ritual practiced by Brahmins for several millennia, Athavale spread it to the masses on a large scale, sometimes calling it an "atom-bomb," i.e., a massive chain reaction. Essentially, it was used as a simple means of reminding oneself of the presence of the Indwelling God three times a day, in the morning, before meals, and at night. Interpreting its nine Sanskrit verses, Athavale emphasized (2000a), "The divine force is always with and within us. He wakes you up in the morning, digests your food, and makes blood, and finally gives you sleep in the night. The supreme power that preserves the universe, that makes stars and planets move, resides in you also and so you are neither poor nor helpless."

Over the course of a few years, he started developing a large following. In addition to *Sandhyā* prayers, he made use of several other traditions such as *Tīrthayātrā* (Pilgrimage), *Ekādaśī* (eleventh day of the month in the traditional Hindu lunar calendar), *Upavāsa* (Fasting), *Yajña* (Fire Sacrifice Ritual), and so forth (Athavale 2000b). He stressed the need of using devotion and pilgrimage for the social uplifting of the masses in addition to performing them as spiritual practices. The "devotional visits" laid down the foundation for mass movement led by Athavale that emphasized his core message – "Brotherhood of humankind under the fatherhood of God."

After the 1970s, he advanced the concept of devoting one's time and striving for spiritual advancement by quoting from the BhG.[9] He inspired thousands of professionals from diverse fields, such as farming, fishing, engineering, medicine, teaching, jewelry making, and the dairy industry, to offer a part of their professional expertise and time for the greater spiritual and social good. He compared this with the ideal of the warrior Arjuna who selflessly dedicated his professional skills in the MBh war. According to Athavale, following Arjuna's example, one should proactively perform one's duty, but should offer the fruit of

---

9 *Yataḥ pravṛttir bhūtānāṃ yena sarvaṃ idaṃ tataṃ. Svakarmaṇā taṃ abhyarcya siddhiṃ vindati mānavaḥ* (BhG 18.46). Literally, by one's own actions, one can attain the perfection. This verse was quoted by Athavale to inspire his followers to devote their time and skill for "god's work."

this activity to the Almighty, thus marrying the ideals of action and non-action. This idea of selfless offering was rendered in his socio-spiritual slogan, "*Bhakti ek sāmājik śakti hai*," i.e., devotion is a social force.

Together these teachings of Athavale became the chief methods of practicing and expanding Swadhyaya in India and later the diaspora communities abroad. Since its inception, it has not built any costly buildings or temples, so it never had to rely on any appeals for donations or membership fees. It devised a modern method to spread Athavale's teachings among his followers by using videotapes. Outside India, its activities are exclusively within the Indian diaspora in the West.

In the newly independent India, some observers also considered the Swadhyaya's social work as a new hope for Indian villages (Paranjape 2005; Sheth 2002). Many young idealists eager to feel that they were doing something to "serve their nation and culture" joined Swadhyaya and led the youth programs such as the *Bāla Saṃskāra Kendra* (culture-teaching centers for children) and *Divine Brain Trust* (cultural groups for teenagers). Swadhyaya also offered new roles to women apart from their routine household chores. In devotional visits to their neighborhoods, women became the key players in building new relationships and managing the *Mahilā Kendras*, cultural centers for women.

Athavale, a householder himself with a wife and daughter, downplayed the role of asceticism and self-mortification as the means of practicing spirituality. Swadhyaya appealed to lay householders since it did not reject the materialistic pursuit of money and family. Athavale's focus on "reason-based religion," with its "rational" interpretation of tradition, appealed to his followers. There was neither a secret mantra to meditate nor a magical formula to be recited. There was less emphasis on "old-style" rituals. Members of the Indian diaspora outside India, concerned about their children's cultural values in a new country, found an ideal solution in the cultural educational classes offered by Swadhyaya. Celebrations of Indian festivals by Swadhyaya have also played a major role in attracting new people to its fold.

To borrow Max Weber's term, Athavale's *charisma* was *routinized* into a loose hierarchy of local and regional Swadhyaya community leaders. They model themselves after Athavale by striving to achieve intense knowledge of Indian texts and Indic traditions. They study the Swadhyaya books that are based on Athavale's discourses on Indic texts and history, and take the annual examinations based on these books. They also dedicate several hours every day to Swadhyaya community activities as a form of devotion to God.

Swadhyaya spread widely in the Indian states of Maharashtra and Gujarat. However, in other states it has not made any noticeable impact. One reason for this could be that Athavale communicated in the local languages, Marathi and Gujarati, in the initial decades of Swadhyaya. Most of the senior leaders of Swadhyaya continue to prefer these languages in communication. The language barrier may be one reason for Swadhyaya's limited spread in other states, although it made its mark in some districts of Madhya Pradesh and Andhra Pradesh in the late 1990s.

Swadhyaya's stronghold has been the rural Indian community. Urban Indians, more alienated from cultural traditions than the villagers, have shown less interest in the work of Swadhyaya. Organized religious phenomena have seen a downward trend in recent years (Altemeyer 2004) and Swadhyaya has had to face the same challenges in urban India. Bryan Wilson notes the contrast between the NRMs of Western countries and those of Japan. According to him, Japanese NRMs have been more involved with social institutions than Western ones (1982: 147). Swadhyaya resembles the Western NRMs more than the Japanese ones in this regard.

With its modest beginning in the 1940s, Swadhyaya has covered a huge ground. Slowly awards and prizes started recognizing Athavale's work. In 1996, the Magsaysay Award was given to Athavale for community leadership.[10] The Indian government and other Indian institutions have showered several awards and prizes on Athavale.[11] Athavale also received the Templeton Prize in 1997.[12] Rooted in the Indian dharmic traditions, the scope and width of Swadhyaya goes much beyond its religious work and this is apparent from the awards that he has received which were given for diverse reasons such as social reform, spreading of native languages, community leadership, and environmentalism to name a few. In December 2000, Athavale chose his adopted daughter Jayashree Talwalkar, *Dīdī*, literally elder sister, as the new Swadhyaya leader.[13] On October 25, 2003, at the age of 83, Athavale passed away in Mumbai. Swadhyaya continues to spread in India and in the Indian diaspora abroad. However, most of its followers still are people from the Western Indian states of Maharashtra and Gujarat. Outside India also, it has remained an Indian organization rather than a cross-cultural transplant (Levitt 2004).

Swadhyaya's Dharmic Environmentalism

I now present Swadhyaya as one of the examples connecting dharma with ecology. Two eminent sociologists, Gordon Melton and Bryon Wilson, have presented theoretical frameworks to study New Religious Movements. I now apply their observations vis-à-vis Swadhyaya to discuss the paradigm of "New Religious Movement." In my analysis, I present several features of Swadhyaya that resemble different world traditions.

---

[10] www.rmaf.org.ph/Awardees/Biography/BiographyAthavalePan.htm (viewed on January 10, 2006).

[11] For a list of awards, www.Dadaji.net/awards.htm (viewed on April 20, 2007).

[12] www.Templetonprize.org/bios_recent.html (viewed on April 20, 2007).

[13] Like other NRMs, this gave rise to several controversies involving its other senior leaders that are beyond the scope of this book. See http://vmehta.conforums3.com/index.cgi?board=Religions.

As A. Whitney Sanford suggests, the general understanding of the term "religion" is largely based on the Western theological connotations with definite scripture, founder, historic events, and specific theologies and practices (2007). Transcending the boundaries of "religion," some Indian thinkers, including Athavale, have included Jesus Christ and Muhammad as the eleventh and twelfth incarnations of Viṣṇu, thus incorporating Christianity and Islam into Indic traditions. This "inclusivism" of Athavale matches with other neo-Hindu thinkers such as Mahatma Gandhi and Swami Vivekananda.[14]

As explained earlier, "dharma" is a better representation of Indic spiritual phenomena than "religion." Swadhyaya also shows less features of a new "religion." On the one hand, Swadhyaya presents a definite set of practices, deities, philosophies, and scriptures as interpreted by Athavale and, on the other hand, it contains open boundaries based on its Indic "dharmic" foundations. There is no conversion or covenant or commandments required of followers of Swadhyaya. One can continue to practice one's theologies and rituals and still participate in Swadhyaya activities. Many New Religions of Japan have offered new models and solutions regarding the salvation of their followers (Wilson 1990). In contrast, Swadhyaya has offered no such promises of salvation. Many New Religions of the past, such as Buddhism and Protestantism, arose in opposition to prevalent social religious elitism. Swadhyaya does not demonstrate any such rebellious qualities either. Thus, it transcends the category of "New Religion."

According to Wilson, another common feature of the New Religious Movements or New Religions is that they transgress the dichotomy between priests and the lay community. They give rise to more egalitarian structures with access to the scriptures for all. He also mentions that many times they tend to reform the parent tradition. Swadhyaya has also shown this feature. Athavale repeatedly used to say that he is not presenting anything new to society, but just renovating and rejuvenating the "original Vedic tradition." Athavale also called his followers "part-time priests," thus rejecting the dichotomy of official

---

[14] Halbfass notes the neo-Hindu trend for the universalistic inclusion of "others" (mlecchas) into the semantic scope of dharma as an influence of Western liberal humanist ideas (1988). However, it can be added that the Buddha was also absorbed into the Hindu pantheon as the ninth avatāra of Viṣṇu (Joshi 1967). Similarly, several other such gods, goddesses, myths, and legends have been included within the dharmic scope much before Western liberal humanism of the Enlightenment period. Halbfass also argues that the inclusivism of neo-Hindu thinkers is based on their philosophy of Advaita Vedanta in which the "Nameless and Formless Transcendental" is the supreme reality and all theistic adoration of form as inferior. However, Athavale saw the formless aspect of the One Reality to be merely the aura of the Supreme Person, whose Transcendental Form is the supreme reality. For Athavale, theistic adoration based on Saguṇa Mūrti Pūjā was superior to transcendental Advaita Vedantic Nirguṇa Brahman. In my opinion, like most European Indologists, Halbfass heavily leans toward Advaita Vedanta for drawing his examples and making his conclusions about neo-Hindu thinkers such as Vivekananda and Radhakrishnan (Biernacki 2007).

priests and lay people. For Athavale, the Vedic period was a golden period in which Indian society had achieved a perfect balance of socio-economic harmony and spiritual quest. For him, Vedic Aryans were proactive in spreading their culture across India and beyond. According to Athavale, Indians in the last 1,000 years, mainly in the medieval period, lost their original traditions and hence suffered the spiritual and socio-economic downfall. It is only by going back to the original Vedic teachings that Indians can bring back the "Vedic Golden Age." While this zeal for going "back to the Golden Age" matches with several other New Religious Movements in Christianity, Islam, and other religions, Athavale's emphasis on devotion and on the Vedic tradition finds itself noticed albeit in a completely different context by Frederick Smith (2006: 596). Smith writes, "[T]he Vedic seers would probably be more at home with any one of a number of devotional and ecstatic sects that arose in India than with the philosophically rigorous and repressive orthodoxies that were established in their name." Like other devotional sects founded or inspired by bhakti saints, Athavale had also emphasized the role of bhakti and devotion in all his discourses and experiments. His entire project presents itself as a simple bhakti oriented movement rather than a philosophically rigorous one. Even the usual philosophically and psychologically emphasized practices of yoga and meditation that are the key practices of other neo-Hindu movements, have found little mention in Athavale's discourses. His simple vernacular message rooted in bhakti is what brings him more in line with the Vedic seers than philosophers or authors of classical traditions of India (Rukmani 1999).

According to Wilson, almost all the New Religions and New Religious Movements in the West have faced strong opposition from members of established traditions. However, Athavale did not face any opposition from other older Hindu leaders or gurus. He did not mix or merge Swadhyaya with older sects or organizations, but still enjoyed the appreciation of Hindu leaders while maintaining the distinctive identity of Swadhyaya. Likewise, older Hindu organizations also never felt threatened or challenged by Swadhyaya; rather they saw Swadhyaya as actively fulfilling their own original vision and mission.[15] Daniel Gold has described Swadhyaya as a "Vedic family" and has differentiated it from other Hindu nationalistic organizations (Srivastava 1998: 172–86). It has its own definite set of Sanskrit prayers and other devotional songs that are sung in gatherings in a prescribed way. It has its own set of images of Hindu deities Kṛṣṇa, Śiva, Gaṇeśa, and Pārvatī that are placed in all its gatherings and programs. It has its own emblem and a flag that is unfurled on special occasions.[16] From the above

---

[15] For instance, Ramakrishna Mission President Late Swami Ranganathananda mentioned in an email to me that Swadhyaya is fulfilling the dream of Swami Vivekananda, December 17, 2004.

[16] I have attended several large gatherings of Swadhyaya in India, the USA, and in Canada where I have noticed distinct set of images, flag, and an emblem of Swadhyaya being displayed on the stages.

description, Swadhyaya appears as a reform movement with open boundaries based on Athavale's interpretation of Vedic traditions. It shows less features of a "New Religion" in which one would seek conversion or membership.

Regarding the importance of the charismatic figure in a New Religion as suggested by Wilson, we could again place Swadhyaya between the two extremes of Indic traditions and Abrahmic religions. While Indic traditions do not have any single prophet, messiah, or founder, Christianity and Islam are based on the charismatic figures of Jesus and Muhammad. Although Swadhyaya is inspired by the charisma of Athavale, it is attributed to his intellectual leadership and innovative methods, in which he reinterpreted traditional Indic ideas and applied them to revive the culture. His charisma was not based on any divine revelations or miraculous experiences. Although Athavale is seen by his followers as a Vedic seer rather than as a miraculous mystic, he has been rendered legendary through the popularization of the events of his early life. For example, his intense studies at the Asiatic Society Library in Mumbai and his photographic memory are highly regarded by his followers. His ancestors are also held in high regard. In general, Athavale is not regarded as a god or *avatāra*, although some dedicated followers might privately regard him as a person with a divine mission or an *anśāvatāra*, literally a partial incarnation. For most followers however, Athavale is a revivalist of Vedic culture and for them the true Vedic culture is what was preached by Athavale. Thus, the perception of his followers is not to see him founding a "New Religion" but bringing back the vision of Vedic sages for the modern times.

According to Wilson, modernity has posed several challenges to old religions. For instance globalization, secularization, and the growth of mass media have significantly affected traditional religions. However, Swadhyaya's early followers came from the villages of Maharashtra and Gujarat that hardly had an impact of modernity in the 1960s and 1970s, so we cannot fully apply Wilson's observations. He writes that NRMs can make use of modern "rational" methods of recruitment, fundraising, publishing, and other organizing activities. However, Swadhyaya refrains from making use of these modern organizational techniques in most of its activities. In Swadhyaya, there are no membership drives and no membership rules, fees, or donation required of new members. Most of its volunteers tend to avoid modern communication means, such as the Internet, for managing various activities. They have also been slow in approaching the media to promote Swadhyaya unlike most modern organizations and NRMs with huge PR campaigns. An exception to the traditional means and practices of Swadhyaya is the use of videotapes of Athavale's discourses. Athavale gave thousands of discourses in Mumbai that were meticulously video-recorded, and these videotapes are now circulated all over the world in Swadhyaya gatherings. John Little has called them "Video Vachana" (1995). Again, unlike most "New Religions," Swadhyaya presents little evidence of using modern means to spread itself as an organization, instead most of its practices are rooted in traditional Indian ways of running a joint family based on personal relationships, as noted by Daniel Gold (Srivastava 1998: 172–86). However, Swadhyaya's rejection of modernity is not a Gandhian argument against machines and consumerism.

Further, Wilson suggests that in today's globalized world, past community links are getting weaker as families are becoming "nuclear" and mobile in search of greener pastures across the globe. Swadhyaya's success outside India can be attributed to this transition as well. Indian immigrants in an adopted new land find community resources in Swadhyaya's family paradigm. The followers of Swadhyaya literally call each other brothers and sisters. By regularly meeting in Sunday sessions, slowly the new immigrants develop new relations with other members of the immigrant community. In this way, Swadhyaya provides an alternative and extended sense of family in the wake of globalization.

Wilson writes that in the Western countries priests and clerics are now paid smaller salaries. The NRMs recruit more lay people for their various services and do not rely on any special class of priests unlike older religions. The Indian situation is a little different from this.

In modern India, priests (Brahmins) have already been rendered economically and politically powerless.[17] Brahmins in the time of Swadhyaya's ascendancy were relatively irrelevant in Indian society and hence Swadhyaya cannot be seen as offering any "liberation" to lower classes from the hold of the upper class. Thus, Swadhyayis cannot be compared with the "liberation theology" movements in Christianity where lower classes made use of "New Religious Movements" for social liberation. Swadhyaya is strongly rooted in Indian myths and legends with high regard for traditional values. Instead of rejecting or rebelling against Brahminical values, they strive to spread Brahminical roles and duties even to so-called lower castes and classes. According to Wilson, New Religious Movements arise when the official clergy fails to connect with the lay community. The rise of New Religions can thus be attributed to the search for alternative paradigms by the lay community. This is true of Buddhism and Jainism in ancient India. Both those religions arose as new forces challenging the Vedic rituals performed by Brahmins. However, Swadhyaya arose in a different setting. When Athavale formally launched Swadhyaya in the 1960s, people were becoming largely "secularized" in cities in newly independent India. Urban Indians had interest in economic progress and little interest in religion. Swadhyaya's limited success in cities can be attributed to the secular, non-religious, and modern atmosphere.

On the other hand, Swadhyaya has been a success in the villages of Maharashtra and Gujarat. Even here, Swadhyaya's spread was largely due to the

---

[17] After India's independence, the so-called backward castes have been privileged in almost every sector of society by the Indian government. In contrast, Brahmins, the so-called upper caste, have no such advantages and their traditional professional job of observing rituals does not offer them any economic benefits either. Politically, Brahmins have been increasingly marginalized as noted by *Outlook*, a leading Indian magazine (June 2007). The percentage of Brahmin lawmakers dropped from 19.91 percent in 1984 to 9.17 percent in 2007 (*Wall Street Journal*, January 2, 2007).

active attempts of Athavale and his early followers to try to connect with the villagers on a regular basis. People were not really looking for an alternative paradigm from their established traditions and thus it was a "top-down" approach rather than a "bottom-up" one. Helen Hardacre identifies some common themes in Japanese religions (1988). Most of the Japanese New Religions have been said to represent reactions to social problems and they include common elements of ancestor worship, shamanism, and faith healing. Again, Swadhyaya does not seem to resemble Japanese New Religions in these ways. It is not a reaction to any social problem. Frequent "devotional visits" undertaken by Athavale and his early followers developed a bond between them and the villagers. In addition, this bond helped spread the doctrines of Swadhyaya. My point is simply that the emergence and spread of Swadhyaya was due to the proactive efforts by Athavale and not by his followers. It was also not a reaction to the established Brahminical or any other religious authority. His doctrines were meant to inspire people to connect to dharmic traditions in novel ways rather than to reject them. It can largely be seen as a reform movement within the mainstream Indian traditions rather than a "New Religion" or as a solution to "old" religion.

Athavale also sought to present his work as a "devotional" work, cutting across the boundaries of religions. In his discourses, he used to quote from the Bible, the Qurān, the Torah, from the lives of Sikh gurus and the Buddha. Many times in his discourses, he also identified Jesus Christ and Muhammad as the eleventh and twelfth incarnations (*Avatāra*) of Viṣṇu. His inclusivistic spirit is similar to that of Gandhi but it is absent in religious sects whose inspiration is based on one Indian text or one Indian guru. While the Hindutva movement is based on Hindu nationalism against Muslims, Swadhyaya stays away from nationalistic jingoism. It has never participated in any of the Hindutva-sponsored campaigns. Pramila Jayapal has noted that Swadhyaya has also played down the word "Hindu" to define its identity in an attempt to appeal to people across all religions (Srivastava 1998: 116–18). Moreover, in some villages, Muslims and Christians participate in Swadhyaya's activities.[18] Since 2007, thousands of North American Swadhyayis celebrate Christmas by donating blood to pay their homage to Jesus Christ (*Chicago Tribune*, December 26, 2007).

Athavale had intensely studied different world religions and philosophies in the early part of his life in the Asiatic Society Library in Mumbai. He used to call Karl Marx a seer because of Marx's concern for the poor and downtrodden. His appreciation for Marx again shows his eagerness to transcend the dichotomy of "religion" and "secular" work. Few Hindu leaders would study the works of Marx, let alone admire it. Similarly, one of the most interesting features of his discourses on the ṚV is his mini-series on Jewish history and mythology. In

---

[18] From a report about Muslims participating in the BhG elocution competition organized by Swadhyaya in *The Times of India*, December 7, 2002. Several observers have noted that Muslims recite from their Qurān while Hindus recite Sanskrit verses in Swadhyaya gatherings in some villages (Srivastava 1998).

reference to one of the Vedic verses about the qualities of a social leader, he mentioned Moses as an ideal social leader and then traced the entire Jewish history starting from Abraham to the modern nation state of Israel. He exhorted Indians to follow the Jewish example, and to have respect for their ancient language Sanskrit, their nation, and for their traditions. He cited the revival of Hebrew as an example in this regard.

As noted by Ananta Giri (2009), Athavale launched several socio-economic experiments based on the theme of devotion. Some of these experiments, such as *Yogeśvara Kṛṣi* (devotional farming) can be compared with collective farming that was implemented in Jewish Kibbutzim, as noted by Pramila Jayapal (Srivastava 1998: 111). Similarly, in some of his discourses he quoted from Islamic writings[19] in support of the Vedantic vision of "Indwelling God." He also showed a deep appreciation for Namāz, the Islamic religious prayer system. In one of his discourses, he cited an example from the Mughal king Aurungzeb's strict discipline of offering Namāz. In 1997, Athavale organized a huge pilgrimage of millions of Swadhyaya followers to the Sikh holy town, Nanded, in Maharashtra, where he was honored inside the holy Gurudwara with traditional Sikh rituals (*Tattwadeep*, December 1997). Swadhyaya has periodically organized conventions of its followers from around the world. One can observe Swadhyayis from the Western countries coming to small Indian towns and staying at the homes of the so-called lower castes or poor families in villages. Majid Rahnema has noted this intermixing of people from different classes, castes, and religions that leads to better social cohesion (Srivastava 1998: 139–53).

Based on Swadhyaya's work with downtrodden and other so-called backward castes, many observers have called Swadhyaya a "social" movement rather than a "religious" one. In 1999, the Indian government awarded Athavale the second highest civilian award, the *Padmavibhuṣaṇa*, for his social reforms. Some observers have compared Athavale's work with Gandhi's social work rather than with that of religious leaders. Several features of Swadhyaya match those of a "secular" NGO than a "religious" one and hence I suggest that Swadhyaya can also be categorized as a "social" movement. Following Gerald Larson (2004) and McKim Marriott (1990), I suggest that the Western categories "New Religious Movement" and "New Religion" seem inappropriate for Swadhyaya. Alternatively, the category "dharma" based on the Gandhian definition (that includes duty, religion, and ethics), appropriately describes multi-faceted activities of Swadhyaya. Swadhyaya presents several features resembling those of a religious movement while other features match those of a social movement. Hence, it may be better to call it a socio-spiritual movement that transgresses the dichotomy of secular and religious worlds. Using the Indic category, we can

---

[19] "*Adam ko khuda mat kaho ādam khuda nahi, par khuda ke noor se ādam juda nahi*," literally, do not regard the man as God but the man is not separated from God's divine spark, quoted from Athavale's discourses.

rather call it a dharmic community based on the concept of "dharma" which incorporates socio-economic and spiritual objectives.

Several of the Swadhyaya prayogs have ecological implications in addition to spiritual and socio-economic ones. According to Jitendrabhai, a senior Swadhyayi volunteer working in Kheda district of Gujarat, Swadhyaya prayogs are essentially done for human transformation with their ecological benefits as mere "by-products." The mission of Swadhyaya is to generate and spread reverence for humans, animals, trees, earth, nature, and the entire universe in general.[20] Swadhyaya deploys several practical experiments to develop reverence for the society and for the ecology. In addition to weekly discourses of Athavale that people are expected to watch as a congregation, they are also expected to participate in different "prayogs" developed by Athavale. As of May 2007, the total number of *Vṛkṣamandiras* (tree-temples) is about two dozen; the number of *Yogeśvara Kṛṣi* (devotional farming sites) is about 10,000, and the number of water harvesting sites is over 100,000. Last *Vṛkṣamandira* was established in Kenedy, Jamnagar in the year 2000 (I have provided the complete list in the appendices).

Based on the devotional teachings of Swadhyaya, Swadhyayis have planted millions of plants and developed about two dozen tree-temples in the Indian states of Gujarat, Maharashtra, Andhra Pradesh, and Madhya Pradesh. Athavale cited several dharma texts to theorize what I call the "arboreal dharma," a dharmic environmentalism to worship and nurture the trees for their unique qualities. Similarly, his other experiment for cows can be called the "bovine dharma," dharmic environmental ethics to serve the cows as mothers. Further, I present Swadhyaya work related to water harvesting and agriculture as the "earth dharma," religious and ecological ethos pertaining to the preservation and sustainable cultivation of the earth. In the appendices, I also present Athavale's thoughts related to other natural entities.

Swadhyaya's Arboreal Dharma[21]

Tree-temples are based on Athavale's teachings of regarding trees as gods. By several explanations from Indic texts, he developed a set of preachings that I would like to term "arboreal dharma," dharmic ecology inspired by the qualities of trees. This is different from Emma Tomalin's generalized statement that

---

20 This is quite similar to Nobel laureate Christian theologian Albert Schweitzer's philosophy of "reverence for life," who was compared with Saint Francis of Assisi for his views on universal reverence (Meyer and Bergel 2002). According to Schweitzer, environmental pollution essentially is a product of mental pollution and if human minds can be transformed by proper thoughts and love, environmental pollution would automatically stop.

21 Information in this section is based on my translations from Swadhyaya's Hindi books *Saṃskṛti Pūjan* and *Eṣa Pantha Etad Karma.*

ecotheologians only offer interpretation of traditions, i.e., reconstruction and reinterpretation of traditions to connect them with environmentalism (2009). Athavale offered *traditional interpretations* of Hindu myths and legends without any major *reinterpretation* or reconstruction. He gave slogans such as "*Vṛkṣa main Vāsudeva*" (literally, Kṛṣṇa in trees) and "*Paudhe main Prabhu*" (literally, God in plants). To explain divine power in trees, he interpreted the capillary action of trees in this way, "There is a divine power in trees which makes it possible for water and fertilizer to rise from the roots below and reach the top portion against the gravitational force. It is not just the result of *Keśākarṣaṇa* (capillary action) but it is *Keśavākarṣaṇa* (Kṛṣṇa's force)." According to Athavale, ancient Indic sages had the spiritual vision to see divinity in the entire universe. Since it is difficult for common people to see this transcendental reality in their routine lives, sages specifically asked them to revere some representative plants. Athavale explained that the sages deliberately chose tulsi, which has no material benefits for humans. It provides neither shelter nor flowers nor fruits, and yet sages considered it sacred. Millions of Indians establish this plant in their homes and worship it. Similarly, bilva is considered sacred for Śiva and even the common grass dūrvā is used in rituals such as Gaṇeśa-worshipping. He also explained the ritual of *Vaṭa-Sāvitrī* performed by married women. They tie the sacred thread, *janeu*, to vaṭa and pīpal trees and worship them. Women desire long lives for their husbands and families from these huge trees. According to Athavale, ancient sages introduced the sacred use of specific plants and trees in all these rituals to develop reverence for them. Evidently, Athavale's teachings are based on the Upaniṣadic philosophy to see divinity in every particle. His emphasis here is to inspire his followers to see divinity beyond their immediate socio-economic needs. Although it is easy to develop respect and reverence for one's own family, the ultimate goal of dharma is to transcend this limited family and see the entire universe, including natural resources such as trees, as a family. This dharmic approach is different from shallow ecology's utilitarian approach, i.e., to protect ecology for human needs. This is also different from deep ecology's biocentric approach of privileging nature more than human society. The dharmic approach is to connect the humans with the ecology based on the divine relationship between the two, not by separating one from the other. The dharmic approach is also different from the religious approach because it is not based on one's belief on the "truth" of the words of a historic person or a scripture.

Athavale explained several ethical qualities that we can learn from the trees. Trees develop a fixed bonding with the land. Trees are forever connected with the land that supports them. They grow at one place and provide shelter to others. They are stable and withstand all natural phenomena such as thunderstorms, cold winters, and hot summers with courage and patience. Athavale compared Śiva, who drank poison so that other gods could get nectar, with the trees which intake carbon dioxide so that others can get oxygen. Trees provide fruits even to those who throw stones at them. They provide roots and herbs for medicines, leaves and flowers for sacred rituals, fruits for our physical strength, timber for our houses, and shelter for travelers. In this way, trees sacrifice all their parts for others without any expectation of gratitude in return. They have no false pride for all the charity they do for others. He said that trees serve as ideal role model for ethics and they are wonderful gifts from God. Here, Athavale describes inherent qualities and virtues of trees, i.e., *the dharma of a tree.* Just by observing and following the dharmic qualities of a tree, one can develop one's moral and ethical qualities, *the dharma of a human being.* Here, we see several meanings of dharma interplaying with each other. The dharma, *inherent quality*, can inspire the dharma, *virtue*, and this can help develop the dharma for the environment, *environmental ethics.*

Athavale then goes on to cite several texts about the above-mentioned dharmic interplay of meanings. According to his interpretation of the BhG (15.1), this world is the aśvattha tree whose leaves are the Vedas.[22] Just as the leaves decorate a tree, Vedic knowledge decorates the world. Athavale also cited Shakespeare's appreciation of nature from *As You Like It*, "Tongues in tree, books in running brooks, sermons in stones, and good in everything [*sic*] ... Under the Greenwood tree, who loves to lie with me? ... Turn his merry note, unto the sweet bird's throat! Come hither, Come hither, Come hither, here shall he see, no enemy, but winter and rough weather." He cited Kālidāsa's *Raghuvaṃśa* to show that the devadār tree, regarded as a son by Śiva, is nurtured by goddess Pārvatī. Also from the same text, when Kautsa, the disciple of Varatantu, visits King Raghu, Raghu asks Kautsa about the well-being of the trees at the hermitage of his guru, "The trees of hermitage that you nurtured like your children, who provided comforting shadow to the travelers, are they safe from thunderstorms and other calamities?" According to Athavale, this conversation shows that the ancient Indians had a sense of kinship with trees and they asked for their well-being. Athavale mentioned other verses from Kālidāsa's *Abhijñānaśākuntalam.* Once while Śakuntalā was watering the plants, her friend Anasuyā tells her that her father Kaṇva seems to have more affection for his trees than she does. Hence, he asked her to water the plants. Śakuntalā replies, "I feed the plants not just to obey my father but I love and feed them considering them my own brothers." Later, when Śakuntalā leaves for her husband's home, Kaṇva asks his trees to bid farewell to her, "One who did not even drink water before feeding you, one who loved to decorate but never took even a leaf from you, one who celebrated when your first flower blossomed, that Śakuntalā is today going to her husband's home, please bid her farewell" (*Abhijñānaśākuntalam* act 4.9).

In another of his discourses, Athavale mentioned the Ramayana legend of Paraśurāma who turns from being a tree-destroyer to tree-protector when he sees Rama watering the Tulsi plant at Panchavati. This is the second meeting of

---

22 *Ûrdhva-mūlaṃ adhaḥ-śākhaṃ aśvatthaṃ prāhuravyayaṃ. chandāṃsi yasya parṇāni yastaṃ veda sa vedavit* (BhG 15.1). Literally, it is said that there is an imperishable banyan tree that has its roots upward and branches downwards. Its leaves are the Vedic hymns. One who knows this tree is the knower of the Vedas.

both after their first meeting at Rama's wedding in which he breaks Shiva's bow. After their second meeting, the destructive weapon of Paraśurāma turns into a constructive instrument and he plants the trees of mangoes, pineapple, betel nut, and jackfruit. According to Athavale, even the name of the mountain in the Konkan region where Paraśurāma lived, shows a change of heart: *Asahyādri* (prefix *asahya* means intolerable) becomes *Sahyādri* (prefix *sahya* means tolerable).

All the above references from different texts again exemplify the dharmic interrelationship of trees with human society. On July 12, 1979,[23] Athavale gave a practical shape to his dharmic ecology, when he inaugurated the first tree-temple at the village of Kalavad in the Rajkot district in Gujarat. It was named *Yājñavalkya Upavan*, an orchard for the Vedic sage Yājñavalkya. There were 6,000 trees planted here. Villagers nurture them throughout the year. They stay at a tree-temple not as gardeners but as devotees. The orchard becomes their temple and nurturing the plants becomes their devotion. The fruits or other products of such orchards are treated as *prasāda*, divine gift. The income generated from the selling of such fruits is either distributed among the needy families or saved for future such prayogs.

On May 28, 2007, I met a senior Swadhyayi volunteer, Nathbhai, at his home in New Jersey. Nathbhai was one of the key participants in the building of the first tree-temple at Rajkot, India, in 1979. Replying to my questions, he did not want to label tree-temples as an ecological project. Instead, he simply included them with other such prayogs such as *Matsyagandhā* for fisher folk and *Yogeśvara Kṛṣi* for farmers. He even denied any significant ecological impact of few such tree-temples against the widespread environmental problems. For him, environmental problems are direct consequence of industries and it has to be dealt at that level. Below I quote his responses in my translation. I first asked him about the reason for this seemingly innovative project by Dādājī. He outrightly rejected any ecological basis for these gardens and told me that Dādājī did not plant these trees for *paryāvaraṇa* (literally, ecology) even though he had taught them the reverence for all creation including mother earth, rivers, trees, and animals based on Upaniṣadic concept *Īśāvāsyam Idam Sarvam* (the entire universe is divine because of omnipresence of supreme divinity). Dādājī taught them that the early humans chose water out of the five great elements, *pancha mahābhūtas*, fire, earth, wind, water, and space, to offer to God. Later, based on their discoveries from surroundings, humans started offering other things such as the flowers, fruits, sandalwood, and *abīr* (red powder). Since water and other natural offerings were created by God, humans wanted to offer what was created

[23] Coincidentally, these were the years when Chipko in UP, Appiko in Karnataka, and Anna Hazare's efforts in Maharashtra were being hailed as successful socio-ecological movements based on people's participation. Athavale included tree-plantation projects into his programs of spiritual activism after the emergence of other socio-ecological movements and soon launched several other environmental prayogs related to water conservation, sanitation, and agriculture.

by humans to show their gratitude towards God. Flowers and fruit are sowed by humans but grown by God. Even further, what can humans offer to God? What is their own that they can offer! Dādājī told them that their own creativity, skill, and efficiency could be a better offering than mere flowers or fruits. Since farming is the skill that farmers and gardeners have, Dādājī devised tree-temples for their offering. He simply chose to build tree-temples to help farmers offer their efficiency and time for spiritual development based on the BhG's message of selfless work, *Karma Yoga*. Initially the tree-temple was named *Upavan*. Later on, to distinguish from other gardens, the name was changed to *Vṛkṣamandira*.

Obviously, as I had hinted earlier, environmentalism was never the motive of the founder of Swadhyaya, nor of the followers. Rather, they strongly emphasize that such prayogs are only for expressing their devotion for the Almighty in a novel way. Instead of usual rituals of offering done in temples, devotion can transform one's perspective toward one's daily work-schedule and this routine labor can be utilized for devotion, which Athavale called *kṛti bhakti*, literally action oriented devotion or devotional action. What is also noteworthy is that this devotion is not merely based on the blind faith on the words of their founder guru. Rather their goal is to apply this devotional labor into their work of farming to develop their bond with the trees and to learn the moral qualities from the trees. Ironically, the more I asked him about the role of such gardens for environment, the more he denied it. According to him, ecological problems have to be solved at the global level, not at the local level. Trees are only a small part of ecology; it was not an issue with Dādājī. A small number of gardens cannot really help the environment. Farmers commonly own much bigger fruit gardens for their business that might have a greater ecological impact. Farmers in general have no awareness for ecology. However, they regard even stones as divine, so trees can also be regarded as divine, especially those in which they do *Prāṇa Pratiṣthā*. This is a ritual ceremony to invoke and establish the divinity in the trees. After many years of participating in the Swadhyaya sessions, people start understanding their relationship with God. True bhakti, devotion, is the basis of Swadhyaya. Today, bhakti is removed from daily life and is limited only to performing rituals in temple. To restore bhakti back in daily life, God's role in daily lives has to be understood. People might turn toward God out of fear or out of materialistic expectations. However, Swadhyaya prayogs are to be practiced to reform this perspective. One should remember God selflessly out of sheer gratitude. Prayogs help Swadhyayis transform their lives and continuous participation in them helps them continuously.

Again, it is obvious that for Swadhyayis, their practice of bhakti based on the interpretations offered by Dādājī is a major motivation to participate in these prayogs. However, what they may be unaware of is a powerful force that they have been able to generate that can be a role model for several other traditional communities. While the governmental efforts also routinely organize tree-plantation projects, the survival rates of such plants is questionable.[24] On the other hand, the sincere devotion of Swadhyayis toward the newly planted and deified plants ensures a 100 percent survival rate in all their tree-temples. This is a striking contrast between two similar efforts. The former, which is done at the political level with only ecological focus, and the latter, which is done at community level with only devotional focus. The success of Swadhyayis' efforts is apparent in the way they conduct such activities. Here is a glimpse extracted from Nathbhai's memoir that he narrated to me. This was in 1978 when the first such Swadhyaya tree-temple was to be built based only on the "devotional labor" of Swadhyayis without any political or outside financial support. The intensity of Nathbhai's inspiration was apparent even after 30 years when he was one of the key volunteers for this project. In 1978, a wasteland of about 50 acres was bought and divided into parts of about 23 and 27 acres. It was a rocky and barren land. Nearby farmers were not hopeful about the project but Dādājī insisted and even located a spot for the well and that indeed turned out to be a great supply of water for the tree-temple. Thousands of Swadhyayis came from all over Saurashtra including districts of Rajkot, Jamnagar, Surendranagar, Bhavnagar, Junagargh, Vairaval, Amreli, and Kutch, on weekly basis. Most of them also had their own personal farms so they had to take some time away from their personal farming. It was not noticed by any media or political leaders. As usual, it was a silent work so there was no question of support or opposition from any section of society. Volunteers brought their own food, bedding, and worked completely selflessly with only the incentives of devotion, selfless love, and selfless work for God. After working for about one year, on July 12, 1979, the tree-temple was inaugurated with *Bāla Taru Prāṇa Pratiṣthā Mahotsav* on the first piece of the land. Next year the second portion of the land was inaugurated. *Bāla Taru*, a small plant, was ritually invoked with God and then was regarded as a deity. A few families from nearby villages and towns were sitting near each bāla taru to perform *pūjā* for couple of hours. In the next months and years, that group took care of their assigned bāla taru, so they became *pujārī* (worshipper) for that bāla taru. Just as a priest maintains and cleans a temple, people did the same for these small plants. Some plants were planted on the fence for decorative purposes. Local coordinators assigned the specific roles for each *pujārīs* working there. Today, almost every district in Gujarat has one such tree-temple. At the inauguration of the first tree-temple, Dādājī reminded the role, *bhumikā*, of volunteers. He told them that they are investing their labor in these tree-temples but there is neither a salary for

---

[24] According to a survey carried out in the 1980s by Food and Agriculture of the United Nations, only 66 percent of the plants planted by state governments survived. The figures are similar for several other countries where the survey was conducted. I also found two reports in *The Times of India* in which serious concerns were expressed about the survival rates of plantation efforts in Bihar and Haryana (2002 and 2003). Also, see a government of India report which admits poor survival rate of forest plantation efforts: http://envfor.nic.in/nfap/forest-plantation.html.

their labor, nor ownership. An outside observer also is neither owner nor a paid employee of these tree-temples. So, what is the difference between an outsider and a Swadhyayi volunteer? He underscored that Swadhyayi volunteers cannot remain *taṭastha* (indifferent or passive) because they are investing their time and efficiency as their devotion so their role is of *pujārī*. They take care of this temple as a *pujārī* and maintain it by performing their pūjā. Only their devotion for these temples distinguishes them from others.

Another Swadhyayi Rajeshbhai narrated a similar incident to me about the *Buddha Vṛkṣamandira* built in Junagarh district. This piece of land was certified by an agriculture officer as a barren land unsuitable for farming. Earlier when the local administration wanted to donate the land to poor people, they rejected the offer, noting its uselessness. Swadhyayis have turned this land into a green oasis in the form of Buddha Vṛkṣamandira spread over 42 acres. It has 1,200 fruit trees, 200 herb trees, and 150 trees of 30 diverse species. To water these plants, a 700-meter canal was also built. The entire orchard also has an underground water drop system about 18 inches under the land, supplying one-liter of water to every plant. This leads to a 70 percent reduction in the wastage of water and 800 plants can be watered simultaneously. This Vṛkṣamandira also has a *prārthanā mandir*, a prayer temple, where approximately 15 *pujārīs* pray and discuss every evening. A charity trust was established and the land was registered in its name. Tree-temple's products are sold in its name and the money thus generated, called *Mahālakṣmī*, divine money, is distributed to the needy families as *prasāda*, divine gift. People are constantly vigilant never to let these temples be used for any selfish use whatsoever. Swadhyaya thoughts are important to maintain the sanctity of the temples.

Shah et al. (1998) record a similar Vṛkṣamandira known as *Śaunaka Vṛkṣamandira* in Rojhad, established on August 11, 1990. Its total area is 11.64 hectares. There are 11,000 eucalyptus trees and 1,617 fruit trees. Chemical fertilizers or chemical pesticides are not used here. Worm compost is prepared, and a mixture of kerosene and extracts from cactus thorns is used as a pesticide; there is a tube well; drip irrigation through a network of PVC pipes has been provided on 14 acres of the Vṛkṣamandira. This Vṛkṣamandira is different from the others – here *pujārīs* from not merely the surrounding villages but from the whole district come to offer pūjā. Approximately 3,690 Swadhyayis (2,443 males and 1,247 females) are currently registered as *pujārīs* at the Vṛkṣamandira. Of them, only 280 *pujārīs* are from the local talukas, the rest come from all over the district; about 300 *pujārīs* cover a distance of more than 100 kilometers from their residence to the Vṛkṣamandira. There are four broad categories of *pujārīs*: (a) *Vānaprastha* couples and individuals (senior citizens) who reside for a period of three months; (b) *Vānaprastha* couples who reside for a period of one month; (c) *pujārīs* who live for a fortnight; and (d) those who spend only a day at the Vṛkṣamandira. The *pujārīs* of the first and second categories are experienced agriculturalists; they are familiar with the routine work of the farm; they guide and help others who may be unfamiliar with the farm work; they act as a link between the daily *pujārīs*. One couple from the first category, two couples and two individuals from the second and third categories, and 10 individuals from the fourth, work as *pujārīs* at the Vṛkṣamandira.

The daily *pujārīs* bring packed meals with them while those who stay for a fortnight or longer cook their meals at the Vṛkṣamandira. All *pujārīs* use their own resources for their transport, food, and other requirements. The daily *pujārīs* come to the Vṛkṣamandira in the evening and leave the next. At night, the *pujārīs* engage in devotional activities and discuss Swadhyaya ideas. The produce of the Vṛkṣamandira is marketed in the area itself. The *pujārīs* themselves buy the produce at a fixed price; some visitors also buy from the Vṛkṣamandira. The income during the last three years has ranged from a little over one-lakh rupees to about 2.5 lakhs. The procedure for handling cash has been streamlined – as soon as the cash crosses the figure of 1,000 rupees, it is deposited in the bank; when the bank balance crosses 10,000 rupees, it is sent to the central office. On the recommendation of the local workers, any proposal for expenditure from the account of the Vṛkṣamandira is forwarded to the central office; the amount is then received and spent for the purpose.

All these descriptions about different Vṛkṣamandiras demonstrate the success of Swadhyaya in convincing the villagers to transcend their reverence from Hindu gods and goddesses to revere plants and trees. Swadhyayis also do not claim that these sites are their efforts to save biodiversity. It is questionable whether the farmers have any such awareness and yet these new "sacred groves" built by Swadhyayis is a small but promising trend to capitalize on the popular devotional attitudes in the service of greater ecological good.

In July 1993, Athavale's followers participated in another prayog called *Mādhava Vṛnda* (Kṛṣṇa orchard). Swadhyayis everywhere planted tree saplings and jointly nurtured their plants for 100 days with daily chanting of *Śrisūktam* and *Nārāyaṇopaniṣad* verses. On October 19, 1993, it was announced that seven million saplings had survived (*Hinduism Today*, July 1995). The tenth anniversary of this devotional undertaking was celebrated in the year 2002 as *Mādhava Vṛnda Daśābdī* (Kṛṣṇa Orchard Decade). These celebrations were held at various places and 3.5 million new plants were planted in the same manner all over India, the USA, and the UK. The perspective for this prayog was to strengthen the feeling that trees and plants are also family members. By periodically chanting Sanskrit verses near plants by the entire family, not only love for plants is manifested but the family members also develop harmony and love among themselves. In some years, the celebration is in form of *Vṛdhiṣu Utsava* when the focus is not planting new trees but taking care of the older ones. According to an estimate by a senior Swadhyayi, there are at least 10 million trees planted so far as part of this prayog. In this way, Swadhyaya's "dharmic ecology" has helped give rise to several grassroot projects to connect local people with ecology. In 1987, Athavale was awarded the "Indira Priyadarshini Award" by the National Wasteland Development Board for tree-temples created under the banner of Vaijnāth Bhāvdarśana Trust. Giving an example of transformation occurring from such prayogs, a Swadhyayi told me that after the Gujarat riots in 2002, in Dhunadra village in Kheda district, they planted trees in Muslim graveyards and Muslims planted in Hindu cremation area. Similarly, another Swadhyayi recalled that Swadhyayi engineers had built sanitation facilities in thousands of tribal villages in Maharashtra.

Overall, we can conclude that the arboreal dharmic work done by Athavale and his followers can be compared with ecological work done by environmental NGOs. However, for the Swadhyayis, their work is simply a reflection of their *kṛti bhakti*, activity inspired by their devotion to the divinity inherent in themselves and in nature around them. For Athavale's followers, trees and plants merely symbolize the divine force that works as a connecting force between the human society, *puruṣa* and nature, *prakṛti*. As Swadhyayis often reminded me: "*To be is to be related*." By developing reverential relationships with the trees, cows, and other ecological resources, Swadhyayis strive to develop their dharmic teachings into practice. Having seen their devotion toward the trees, let us now turn to similar projects that they continue to develop with the cattle, the water, and the earth.

Swadhyaya's Bovine Dharma[25]

By several explanations from Indic texts, Athavale developed another set of preaching that I call "bovine dharma," preaching based on the qualities of cows that emerged as *gorasa*, a project to nurture the cows. This is not as widely implemented in Swadhyaya as tree-temples and other prayogs because most of the followers of Swadhyaya come from the background of farming and fishing. Athavale sought to combine the existing skills of the people with his teachings rather than to ask the people to take on a completely different kind of work in which their skills may not be useful.

Athavale cited a Sanskrit verse to show that seven forces sustain the earth: cows, Brahmins, Vedas, Satis (noble women), truthful people, charitable people, and people without lust and greed. The cow gives all her belongings to humans. Her milk and other dairy products strengthen us. Bullocks are utilized in farming. Cow dung is utilized as a fertilizer. Its urine is used as an Ayurvedic medicine. After its death, its bones are utilized in the sugar industry, skin is used in the leather industry, and the horns are used to make combs. According to Athavale, humans should be eternally grateful to cows. According to him, Indians did not just exploit the cows for materialistic benefits but instead regarded her as a mother. Only humans drink the milk of other species such as cows. Regarding cows as mothers is to express our gratitude and respect for them. Like our biological mothers, they nourish us with their milk. According to Athavale, Indian villagers used to feed cows with love and respect before taking their own

---

25 Information in this section is based on Swadhyaya's Hindi books such as *Saṃskṛti Pūjan* and *Eṣa Pantha Etad Karma* (with my translations).

meals. Even today, many families observe the tradition of *go-grāsa*, offering of symbolic food to cows before meals. Athavale mentioned Kṛṣṇa's love for cows and said that Kṛṣṇa turned into *Gopālkṛṣṇa* (Kṛṣṇa, the cow caretaker) out of his love for cows. Kṛṣṇa used to attract cows with his sweet flute in Gokul. He cited a Sanskrit verse from Padma Purāṇa to show the importance of cows in ancient India, "Let cows be ahead of me, back of me, inside my heart. I should reside in the midst of cows." Similarly, the Brahma Vaivarta Purāṇa mentions that all the gods and pilgrimages reside in the cows.

Athavale explained that the cow has a quality of chewing her food thoroughly before swallowing it. He interpreted it to preach that we should also chew every new thought before accepting it. Only after carefully analyzing it, should we accept it. Such carefully accepted thoughts will not only help our own intellectual development but will also benefit greater society. Extending the bovine dharmic discourse further, Athavale cited the famous BhP verse in which all the Upaniṣadas are called cows, Kṛṣṇa as the milkman supplying milk to Arjuna. We should always be nourished by the teachings of the Upaniṣadas and the BhG.

He also cited another verse by Kālidāsa and extended the meaning of the Sanskrit word *go*. *Go* has several meanings such as cow, physical power, eyes, and speech. "Without go-rasa (milk) there is no taste in food, without go-rasa (physical power), there is no importance of a king, without go-rasa (beautiful eyes), there is no beauty of a woman, without go-rasa (speech), there is no significance of Brāhmins." He preached cow-worship in all its different meanings, i.e., we should respect cow (animals), the rulers should be powerful, women should possess long-term vision, and scholars should convey good thoughts to the masses. Any kind of ritual is incomplete without *Pancāmṛta*, consisting of three dairy products, milk, yogurt, and ghee, in addition to honey and sugar. Milk signifies purity. Our life, character, reputation, actions, mind, and heart, all should be pure and untainted. Yogurt is the second ingredient of Panchāmṛta. Describing yogurt's qualities, Athavale explained that just a small quantity of yogurt transforms a big quantity of milk. Noble people should also develop this quality. They may be small in numbers but they can transform huge sections of societies. The third ingredient is ghee with its unique lubricant quality. Our life should also be lubricated with love for everybody. Obviously, Athavale had very skillfully utilized several metaphors, analogies, myths, and legends to inspire his followers to practice what he preached. In the fifteenth century, Bishnoi guru Jambheśvara also made use of several Hindu stories and other examples to inspire his followers in similar ways.

Like tree-temples, Athavale developed a project for cows called *Gorasa*. He established dairies in villages where people could get the milk throughout the year at a nominal price and the profit earned from such collective effort was distributed to needy local families or saved for future projects. This stopped the earlier practice of selling the milk to far-away cities. Since the intermediaries involved in selling were eliminated, the purity of milk was ensured, and the cost was minimized. This also inspired farmers to domesticate more cows and thus inspired more love and care for animals.

As in the case of his "arboreal dharma" prayog, he sought to develop what I call a "bovine dharma," dharmic ecology for the cows inspired by the cows. Athavale describes the inherent qualities and virtues of the cows, i.e., *the dharma of the cows*. By observing and following the dharmic qualities of a cow, one can develop one's moral and ethical qualities, *the dharma of a human being*. Here again, we see several meanings of dharma interplaying with each other. The dharma, *inherent quality*, can inspire the dharma, *virtue*, and this can help develop the dharma to care for the animals and environment, *environmental ethics*.

Swadhyaya's Earth Dharma

Let me now describe Athavale's thoughts on *Bhumi Pūjan* that I have termed as the "earth dharma." Athavale has inspired tens of millions of Swadhyayis to recite this verse every morning,[26] "*Samudravasane devī, parvata stanamaṇḍale. Viṣṇupatni! Namastubhyaṃ pādasparśaṃ kṣamasva me!*" (O ocean-clad goddess earth, with mountains as your nurturing-breasts. O wife of Viṣṇu! I bow to you and ask for forgiveness as I touch you with my feet[27]). Explaining this verse, he teaches that the earth is the mother of every creature providing shelter to all living creatures and dead substances. She takes care of everyone by feeding her love-nectar. She holds such great attraction that one gets satisfaction only after returning to her after taking a flight into the sky. Scientists appropriately name this motherly attraction as gravitation. Only someone with *gravity* can *attract* someone else. Mother earth possesses this gravity based on her unique qualities such as patience, forgiveness, tolerance, and humility. Continuing his ethical teaching combining both the earth and the ocean, Athavale cited the above verse and called the ocean as her clothes. Just as humans cover most of their bodies with the clothes, the majority of earth is also covered with ocean. Just as the ocean maintains its humility by not crossing its boundaries, humans should also observe humility in wearing clothes and should not wear them with obscenity. Mother earth is a living testimony of kindness. She gives all the precious jewels to her children. She also nurtures her children with her mountains. Since Sanskrit word *payas* means both water and milk, mountains are compared with

[26] Traditional Indians have been reciting these verses in addition to some others as part of their Prātah Smaraṇa, remembering the divine as one wakes up in the morning. Athavale has included some of the traditional Sanskrit verses in his foundational Swadhyaya prayog called Trikāl Sandhyā, chanting of verses thrice a day, in the morning, before meals, and at bedtime.

[27] In Indian culture, it is one of the most disrespectful gestures to touch someone with your feet. It is common to see someone accidentally touching a person or a book or food with their feet and immediately asking for forgiveness.

the nurturing-breasts and are referred to as *payodhara* (carriers of payas). The rivers originating from the mountains nurture the humans just as mothers feed their babies with their milk. We find some stories in the Purāṇas, in which rivers were full of milk. Mountains also carry numerous herbs and vegetables to cure and nourish humans. Mother earth's kindness, tolerance, and mercy are also evident by the process of farming. We dig the earth but get water in return to quench our thirst; we plough the land but get grain in return to satisfy our hunger. Indeed, mother earth is the best teacher to teach benevolence in return for malevolence! She tolerates all our mischief, we run on it, attack her with our feet, build skyscrapers on it, but she tolerates our entire burden with a smile! Hence, it is our gratitude and humaneness that we express by bowing to her. Since mother earth is the wife of Viṣṇu, he is our father. Although the mother serves us, she is not for exploitation rather she is to be revered. Our perspective toward the earth should be of reverence in addition to consumption. As long as we do not understand that only God is the owner of the earth, not humans, we cannot become true children of her. Today, we have started exploiting the earth only for our selfish and egoistic reasons and hence she is becoming barren. There is a story in the BhP that *śeṣanāga*, the great serpent had stationed the earth on its head. When asked whether the weight of the earth, its mountains, and trees bother him, the serpent replied, "*Na bhumiparvatānām ca na me bhāro vanaspate. Viṣṇubhaktir vihīnasya tasya bhāro sadā māma!*" (I do not feel the burden of the earth, its mountains or its trees but my only burden is humans without devotion for Viṣṇu). Athavale explained that a human who does not revere mother earth or Viṣṇu is indeed a burden on the earth. Mother earth has kept her treasure open for us in the form of numerous minerals such as gold, silver, iron, copper, etc. We should cultivate proper emotions to accept her gifts as a divine prasāda and should revere her gratefully. Athavale noted other Sanskrit verses, "*Vasundharā puṇyavatī ca yena!*" (By whom mother earth becomes auspicious) and "*dhanyās tadanga rājasa mālinī bhavanti!*" He differentiated great sages and holy men, by whose touch mother earth becomes sacred, and exhorted Swadhyayis to try to follow such great people. One can become a true child of mother earth by cultivating great virtues and qualities following the great people. When Brahmins change their sacred thread, they touch the sand to their forehead and recite *mṛittike hanme pāpam,* "Sand! Remove my sins." Athavale also quoted Khalil Gibran, "In the heart of the earth, a poem originated and she cultivated trees and gardens in her leisure. We cleared them to maintain our hollowness" (my translation). The earth also has a special fragrance. In the BhG, Kṛṣṇa says, "*tatra sā gandhavatī pṛthvī!*" (There I am in the special fragrance of the earth). "*Puṇyo gandhah pṛthivyām ca!*" (BhG 7.9), literally, the fragrance in the earth is auspicious. Mother earth is warm in the summer and when the rainwater falls upon it, it spreads a unique fragrance. According to Athavale, mother earth has different fragrances in different seasons. Like the earthen fragrance, our lives should also be fragrant with good virtues and qualities. In the ṚV, Śrisūktam's second mantra reveres the earth with several adjectives. The first adjective given to the earth is *gandhavatī*, i.e., fragrant. The second one is *durādharśā*, i.e., that which cannot be taken away by anyone. The land is for everyone and no single person can own or carry it away. The third one is *nityapuṣṭām*, i.e., ever nourishing. Mother earth not only nourishes our bodies with different kinds of food, but also nourishes our minds and hearts. Indian farmers have such reverence for the earth that after ploughing they do not even allow wearing shoes into their farms. Mother earth is ever nourishing for such people who respect and revere her. Mother earth nourishes people living close to her in villages and forests by her "breathings." People in cities have become aloof from her and therefore they do not get her proper nourishment. The next adjective for the earth is *kariṣiṇim*, i.e., that which has abundant dry cowdung. This has an implicit appreciation for cows. Even the Vedas mention cowdung as a natural fertilizer. The next adjective for the earth is *iśvarī* meaning capable and magnificent. Capable and efficient people should protect the beauty and grandeur of mother earth.

Athavale's reverential discourses about mother earth took a constructive shape when his followers launched several projects related to groundwater. In a country like India, where life is dependent on rainfall in large parts of the country, a lot of work is being done to raise public awareness of rainwater harvesting. Centers have been set up in places like Meerut and Chennai. According to the Centre for Science and Environment at New Delhi:

> Our ancestors harvested rain just as naturally as they tilled the ground to grow crops. We lost touch with these local solutions. Now, as the taps dry up, more and more people are reviving this age-old system and practicing it very successfully. Water has been harvested in India since antiquity, with our ancestors perfecting the art of water management. Many water harvesting structures and water conveyance systems specific to the eco-regions and culture had been developed.

Although various Indian organizations have started harvesting the rainwater, Swadhyaya's work in this regard is different because of its underlying inspiration based on devotional reverence for the water and for the earth (Sharma 2000). While inaugurating one of the tree-temples, Athavale had once said, "If you quench the thirst of Mother Earth, she will quench yours." This "earth dharma" became the driving force for Swadhyaya, especially after three successive droughts in 1985–7 in the Saurashtra region of Gujarat. Some Swadhyayi villages started trying out well-recharge experiments. The first such experiment was done at Chokli village, and by 1992 it started taking the shape of a movement and all the nearby farmers began collecting as much rainfall as they could on their fields and in their villages and canalized it to a recharge source. Between 1992 and 1996, 99,355 wells were recharged (*Bhugarbh Jal Sanchay*) and 554 *Nirmal Nīrs* (farm ponds for recharge) were built.[28] According to K.K. Khakkhar, at the

[28] http://laetusinpraesens.org/docs/indiaoz.php (viewed on June 5, 2007).

Saurashtra University, the water recharge activities by Swadhyaya have resulted in approximately 39 million rupees more income. According to Tushaar Shah (2003), Swadhyayis built over 125,000 wells and over 1,000 farm ponds during 1997. Shah believes that if half a million wells in Saurashtra are recharged, the region can solve its irrigation as well as drinking water problem; and at the rate at which the movement is growing, it would not be surprising if this target has been met. The Swadhyaya "water ethic" is, "rain falling on your roof stays in your house; rain falling in your field, stays in your field; rain falling in your village, stays in your village." By 1993–4, the well-recharge movement had begun spreading to other parts of Gujarat, notably Sabarkantha, Mehsana, and Panchamahal districts. Shah laments that the mainstream discourse has largely ignored the impact of Swadhyaya work (2003):

> Some reject it as a religious, sectarian affair; others see these successes as inimitable and donors are lukewarm to it because the movement accepts neither external funding nor can they be catalyzed by funding support; and scholars are stymied by it because it does not fit into the normal linear logic pattern. All in all, then, the larger implications and lessons that the Swadhyaya movement offers in mobilizing masses for regenerating natural resources are more or less lost.

I agree with Shah on his frustration. Swadhyaya's work does demonstrate a great potential to mobilize the community at the grassroot level by invoking the cultural and dharmic paradigms. I hope this study will bring to light the underground work done by Swadhyaya that has largely escaped the attention of scholars. On the one hand, we have a community of religious scholars in the West often struggling to connect ecological concerns with their respective communities. On the other hand, we have examples such as Swadhyaya, which have silently recharged ecological resources based on the religious inspiration of its practitioners.

One of the most remarkable features of this work is its consistent denial of any outside financial, political, or social help. It has continued to depend solely on the devotional inspiration of its own volunteers. According to Jitendrabhai, a senior Swadhyayi, who has supervised several Nirmal Nīr projects in Gujarat in his volunteer work for Swadhyaya for more than two decades, it is important to note the dharmic perspective in Nirmal Nīr prayog. In the morning, Swadhyayis ask for forgiveness since they touch mother earth with their feet as they wake up. They bow to her and call her as the wife of Viṣṇu, thus the earth is the divine mother. Just as Śiva Linga is worshipped by pouring clean water on it, Nirmal Nīr is also their way to worship the earth with clean water. It is an *abhiṣeka*, ritual of sprinkling, done for mother earth. In this prayog, water is cleaned in different stages using sand, brick-powder, and coal and this clean water is stored in the wells and ponds. Water enters the wells through the small holes and splinters in their walls. Existing ponds are also deepened in this prayog. People work together to deepen the wells that also deepens their mutual relationships.

Jitendrabhai also recalled the visit of Magsaysay Award winner water conservationist Rajendra Singh to the Nirmal Nīr prayogs. Rajendra Singh was amazed to see that most of the Nirmal Nīr ponds and wells were built by Swadhyayi people working not in their own villages but in different villages completely selflessly. Thousands of people are motivated to build ponds and wells for other villages because these prayogs are not for selfish reasons. They are not merely out of the necessity of water but they are essentially tokens of gratitude for mother earth. While Rajendra Singh's Tarun Bharat Sangh had to work hard to motivate people of the same village to build ponds for themselves, Swadhyayis of one village worked for other villages selflessly with great motivation. Everybody brings his or her own food and shares it with others. This working and sharing with each other builds and develops relationships among all castes and classes of the village. Muslims work together with Hindus, Harijans work together with Brahmins. Muslims recite their Qurān verses while Hindus recite Sanskrit verses at the time of regular prayers. After the Gujarat riots, Muslims of Lunāwādā village in Anand district had told the district collector that only Swadhyayis should be trusted to deliver the government aid to the Muslim victims of the riots since most other Hindus had turned against Muslims. Earlier farmers used to quarrel if water from neighboring farms would enter their own farms but after working together in Swadhyaya prayogs they now welcome water drained out from neighboring farms to come to their farms. This reflects increased cooperation and unity in the farmers. In another example, farmers of Maganpura village share each other's losses if anybody's cotton crops fail to yield sufficiently. Another benefit of Swadhyaya prayogs that villagers experienced is that the water tables of all nearby wells increased.

The above ideas expressed by a lay Swadhyayi clearly show the zeal and the spirit of the Swadhyaya work. While communal tensions have flared up several times in Gujarat, Swadhyayis show a different glimpse from this state. In addition, anybody who has some experience of social work, such as Rajendra Singh, would be surprised to note the selfless enthusiasm of Swadhyayis. As I noted in the tree plantation work by Swadhyayis, the water harvesting work also highlights that their motive is not just to do the utilitarian work for the village or even to protect the ecological resources of the village. Rather, the underlying objective of Swadhyayis is to bring about constructive social manifestation of their spiritual understanding toward the divine that is dwelling in them, in their village society, and the village ecology. Indeed, Jitendrabhai succinctly summarized this in his own words. In ancient times, Indian sages also were aware about agriculture and ecological resources. For instance, Kaśyapa was referred to as "father of Indian soil." Similarly, Pārāshar is attributed of having written Kṛṣi Pārāshar, a Sanskrit text for agriculture. Similarly, several community leaders and wealthy people had built water tanks called *Bāvadi* in many villages and towns with Hindu idols near them. However, Swadhyaya prayogs are based on Karma Yoga, not Karma

Kānda. Nirmal Nīr prayogs do not have any ritualistic idol worship. They are prayogs for implementing devotion in one's behavior instead of limiting it into rituals. This openness invites people of all castes, sects, and religions to come and work together.

Another similar prayog is called *Sosa Khaḍḍā* (soak pit); this is meant to absorb the sewage water of individual houses outside the house. Sewage water and sanitation is a big problem in Indian villages and cities alike. Drainage is neither built properly nor kept clean by local municipal officers. Swadhyaya has devised a novel prayog for this problem. *Sosa Khaḍḍā*, absorbing pit, is built outside the house in which the water that was originally taken from the earth for use by people is returned to the earth after household use. This avoids the accumulation of dirty water in the villages and prevents the spread of diseases like malaria and cholera. Villagers are told that just as temple of God is kept clean for worshipping in it, every house and every village is also a temple of omnipresent God and hence should be kept clean. When somebody falls in the dirty water, God inside that person also suffers with him or her. Therefore, dirty water should not be allowed to accumulate in surrounding neighborhoods of the village. Near every Sosa Khaḍḍā, a Tulasi plant is also planted. On the rooftop, clean water is made available for the birds. In this way, every family tries to practice reverence for house, village, plants, and birds.

Overall, Swadhyayis' attitudes toward the water and the earth show their long-standing relationships with their ecology which is revitalized by the thoughts and teachings by Athavale. As I will show later, similar ecological perspectives are manifested in the Bishnoi community. I now present the agricultural work undertaken by Swadhyayis that is one of the most widespread experiments in Gujarat and Maharashtra because most of the followers come from the farming communities.

Yogeśvara Kṛṣi[29]

According to Athavale, agriculture includes four components: trust in neighbor, love for animals, faith in God, and respect for nature. Obviously, the second and the fourth components affect ecology directly even though the main activity is about growing grains and not about growing big trees. On festival days such as *akṣaya tṛtīyā* and *dhan teras*, oxen used in Yogeśvara Kṛṣi are bathed and beautifully decorated and their procession is arranged in the entire village.[30] Ladies smear the foreheads of their husbands and oxen with red *tilak* and welcome them. Later, the farms are tilled together where all the oxen participate. Often, the response is so overwhelming that more oxen turn up at a particular piece of land than the

29 Information in this section is based on my interviews with several Swadhyayis.

30 Ann Gold (2001: 128–9) has observed that such celebration of cattle is linked with the collective management of rainmaking and fertility themes. It also helps develop appropriate harmonies among people and between people and the environment.

required number for farming. Thus, this prayog also develops unity, harmony, and goodwill in the entire village. A related prayog *Śridarśanam* involves 20 villages. Twenty people come from different villages and live together periodically on a monthly and yearly basis. For every 20 people, one cow also lives with them for their dairy needs. This cow is dearly cared for by these 20 people. Since this cow is treated reverentially by the people participating in this devotional prayog, their attitude becomes reverential for their own personal domestic animals also when they return home for performing their routine farming. Thus, in this mechanical age of tractors and other machines, the relationship of farmers with their farm animals is strengthened and revitalized. Yogeśvara Kṛṣi also inspires farmers to develop respect and love for nature. One of their slogans is *hariyālī main hari*, literally God in greenery. One of the Swadhyayi volunteers explained that one who goes to a farm just for work is a farmer, but one who goes to enjoy and respect greenery, goes with reverence and gratitude for God. This reverential perspective inspires to make the entire world green.

Another volunteer explained that they cannot avoid the bare minimum violence required during the harvest but they try to avoid harming the nearby plants, trees, and bushes. Moreover, the earlier use of pesticides is now replaced with the herbal powders that are sprinkled on the crops. Vermiculture compost has also been taken up by the villagers on a large scale (Sinha 1998). The products generated from this devotional farming are treated as the prasāda from God. This helps develop the understanding that nature is not just for anybody's consumption or exploitation but it is for reverence since it is a creation by God. Similarly, the inherent love for animals in farmers is strengthened by such prayogs. To recount an example from Yogeśvara Kṛṣi, a Swadhyayi narrated this incident from Timberwa village in the Vadodara district of Gujarat. Swadhyayis go to offer their efficiency at Śridarśanam periodically, as I noted above. When one of them got his turn, his wife was extremely sick and his crop was ready to be cut. He was in a dilemma whether to go or not. Eventually, the couple decided that the husband should give priority to "god's work" and the wife stays alone at home in her sick health. When the husband came back, he found that his farming needs were already taken care of. Later, they discovered that other Swadhyayis had secretly helped the couple by neatly collecting their crops in their warehouse. These secret helpers had criminal records but now with Swadhyaya prayogs, they were so deeply transformed that they were secretly helping others with the feeling of divine unity.

It is important to note the Swadhyaya agricultural prayogs are different from purely "natural farming" pioneered by Masanobu Fukuoka in Japan in the 1940s (Fukuoka 1978). Although natural farming is also being promoted by Mohan Shankar Deshpande[31] and other Indians, Swadhyaya's approach is different. In the latter case, there is no mention of *Kriśipārāśara*, the ancient Sanskrit text about farming. Unlike natural farming or "Rishi Krishi," Swadhyayis do not

[31] www.rishi-krishi.com/index.htm (viewed on January 20, 2007).

hesitate in tilling the ground or using natural fertilizers. Rather than rejecting the commonly used tools and methods of agriculture, the Swadhyaya approach is simply to bring about the ethical, spiritual, and social transformation of the farmers (and of the rest of the society). This transformation seeks to develop a familial interrelationship among the farmers of a small village and a reverential perspective towards the mother earth.

Concluding Remarks

Anil Agarwal (Chapple and Tucker 2000) had mentioned that Hindu beliefs, values, and practices, built on a "utilitarian conservationism" rather than "protectionist conservationism," could play an important role in restoring a balance between environmental conservation and economic growth. Swadhyaya prayogs do not fall in either category. In fact, when I interviewed some of the people who were involved in the first such prayog of tree-temples in Gujarat they vehemently denied both utilitarian and protectionist motives behind their prayogs and underscored the devotional motive instead, as I noted above.

Concurring with Haberman's theoretical framework and his observations from fieldwork on the banks of the Yamuna, these Swadhyaya prayogs are Indian counterparts of what could be called "environmental activism." Similar to Haberman's examples, Swadhyaya prayogs are inspired by Indic traditions. I visited one tree-temple in Valsad in Gujarat in 2006 where local Swadhyayis had told me that Vṛkṣamandiras inspired the farmers to grow more trees even in their personal farms. They also started using more organic and traditional fertilizers, such as earthworms. Their perspective toward trees was changed from exploitative to reverential. When I asked them about the practical challenges or difficulties related to Swadhyaya prayogs, they noted several. The biggest challenge is to be able to sustain the transformation based on the Swadhyaya's teachings. Without the *dharmic* perspective, the work can become "mechanical" or can take the form of another "religious ritual." If the tree-temples fail to inspire people to develop an ethos, develop a bond with the trees, or if Swadhyayis stop practicing this ethic in their daily life, then the work can take a "religious" shape based on the devotional faith on the words of Athavale. That would reduce the dharmic work into another religious ritual as has happened with some rituals in both India and elsewhere. Another challenge is to take these prayogs and replicate them on a larger scale. So far, these have remained smaller local role models at the district level rather than projects at the regional or national level. They also confessed that the number of volunteers available to work at different tree-temples varies according to the intensity and depth of Swadhyaya's thoughts in the surrounding villages. Since the spread of Swadhyaya is not uniform across the different villages and towns of Gujarat and Maharashtra, the number of volunteers working at such prayogs is also not uniform.

Noted environmentalist Anupam Mishra aptly remarked (1993):

> Even without involving the environmentalists, people are bringing out miracles at the grass-root level. Upon seeing them, we should humbly accept them. Even if they may not fit our measuring scale, may be our measuring scale itself may be inappropriate. A work that has already reached millions belongs to the people. Media reports only political parties but it cannot represent the people.

Athavale had developed several more "prayogs" to fructify his mission for socio-spiritual transformation based on the dharmic philosophy of "Indwelling God." I have described only some of them that relate to natural resources such as the trees, the cows, and the earth. These prayogs do not label themselves as "environmental projects" and yet they have succeeded in sustaining natural resources in thousands of Indian villages. Like any other such work, the challenge now is to maintain them and to develop new such prayogs especially now that Athavale has passed away. To my knowledge, the current leadership has not developed new ecological prayogs. However, it seems focused to strengthen the existing prayogs by inspiring more villagers to join the movement. I agree with Ramachandra Guha's remark that there was no environmentalism before industrialization; there were only the elements of an environmental sensibility. The traditional societies did not transcend their locality to offer any systematic vision of reorganizing nature (2006: 6).

The Swadhyaya followers also show similar sensibilities in their local prayogs in the villages. This sensibility in turn is inspired by their cosmology based on the texts, myths, and legends derived from the dharmic traditions. We do see a reflection of textual reverence for nature in the behavior of Swadhyayis. Whether this behavior will take a generic ecological ethos across the time-space limitation is yet to be seen. It is still a nascent movement with the charisma of the founder still fresh in the memories of its followers. Will they become environmentalists that are more active in the new century in the absence of Athavale? Swadhyaya is also emerging as a movement spread around the globe. When Swadhyayis migrate to different parts of the world, will they connect their environmental sensibility to respond to the problems of climate change?

While the evolution of this movement's sensibility into comprehensive ethos in future is still a matter of speculation, I now turn to a different phenomenon that arose in the sand dunes of Rajasthan in the fifteenth century. Bishnois had caught my attention when they first came to the limelight in the case against Salman Khan, a film star, whom they had caught hunting a blackbuck near Jodhpur, the second biggest town of Rajasthan.

Chapter 4
The Bishnoi Community

Heads lost, trees saved, consider it a good deal! It is better to sacrifice your head to save a tree. Accepting price for your sacrifice becomes a stigma on your sacrifice.

Above is my translation of a popular saying among Bishnois of Rajasthan, the desert state of India. In this chapter, I argue that Bishnois demonstrate an overlap of religious, personal, and ecological attitudes. "Dharma" is a term used interchangeably both by the founder and the followers of the Bishnoi community. The legends preserved by their poets and hagiographers for several centuries continue to inspire Bishnois to protect the trees and animals even at the cost of their personal lives. Most of them are not aware of the Western scientific discourse about "global warming" or "biodiversity." For them, their tradition, based on the words and life of their guru, is the main reason for their environmentalist activism. This is much beyond the recognition of bio-divinity based on the Hindu cosmology or Hindu texts. Unlike other Hindu communities, the dharma of Bishnois is not limited just to the Hindu scriptures or rituals but also includes natural resources, as is evident from the examples of their activism that I survey in this chapter. Here is a glimpse of the most well known example about the Bishnois.

It is Tuesday September 9 in the year 1730 in Khejarli, a small village near Jodhpur, a prominent city in the Indian desert state of Rajasthan. Soldiers of Maharaja Abhay Singh, the ruler of Jodhpur, arrive and start cutting khejari trees to be used as firewood for a royal construction project. Amrita Devi, a Bishnoi woman in this village, quickly notices and rushes to stop the soldiers. The soldiers reject her request and continue their plunder. She (and her family) immediately hugs these trees and the soldiers eventually massacre them. The news spreads like a wildfire in the surrounding areas and as many as 363 Bishnois cling to the trees and are killed one by one before the news reaches the king and he stops his soldiers. The question naturally arises: who were these Bishnois, the "tree-huggers," and what motivated them to sacrifice their lives to protect the khejari trees? In this chapter, I endeavor to answer these questions.

Although the Bishnoi community is distributed in several north Indian states, it is most densely located in Western Rajasthan, where their founder Guru Jambheśvara was born. The above quote is attributed to him and is displayed at several Bishnoi temples and villages. The name Bishnoi means the people of the 29 rules (*Bish* and *Noi*, literally 20 and nine).[1] Some Hindi authors such as Hiralal Maheshwari have referred to this community as Vishnoi, the followers of Viṣṇu. This latter name suggests the Hinduization on this (and several other communities) after the partition of India and Pakistan in 1947 (Khan 2004). However, Bishnois themselves insist that the word Bishnoi is based on the Rajasthani words for 20 (Bish) and nine (Noi), representing the 29 rules given by their guru and for this reason they did not approve "Vishnoi" as a replacement for Bishnoi as was done by Maheshwari. *Nagaur District Gazetteer* (1975) also mentions the connection between the 29 rules and the term "Bishnoi" although the statements of Jambheśvara themselves only mention about Viṣṇu and do not mention about the 29 rules (Landry 1990: 11).

The Chipko movement of Uttaranchal is well known for its tree-hugging campaign to resist the tree felling. Its leader Sunderlal Bahuguna and writers such as Vandana Shiva (1988) have speculated that Chipko derived its inspiration from the Bishnois who are famous for sacrificing several lives for similar reasons (Kaplan et al. 1997). Despite this presumed connection of Bishnois with the Chipko movement, little is known about Bishnois' religious and ecological life. Several prominent Indian ecologists have largely ignored Bishnois except few passing remarks[2] (for instance, Landry 1990 and Lal 2005).

To fill this gap about one of the earliest communities with noticeable ecological practices, I began my fieldwork in the villages of Bikaner, Nagore, Jodhpur, and Pali districts of Western Rajasthan. The information I present below is based on my interviews with several Bishnois whom I met in Rajasthan, and my continued electronic conversations with some of them. Some of the information is also extracted from the books and manuscripts that I gathered from the temples and

---

[1] This community is also called the Prahlādapanthi community based on their reverence for Prahlāda, the mythical son of Hiraṇyakaśyapu in Hindu Puraṇas who had invoked the Narasiṃha incarnation of Viṣṇu.

[2] Although Ann Gold (Gold and Gujar 2002: 249) has briefly mentioned about the Bishnois, she also acknowledged the lack of material about Bishnois and encouraged me to research about them. The Indian environmentalists Ramachandra Guha, Madhav Gadgil, and O.P. Dwivedi have not included Bishnois in their various works on Indian environmentalism. Books on Hinduism and ecology also conspicuously lack the accounts of Bishnois. This is somewhat compensated by several media reports that continue to highlight Bishnois, e.g., see "Sect in India Guards Desert Wildlife", in the *New York Times*, February 2, 1993. Other scholarly references that I located are Vinay Lal's article in *The Encyclopedia of Religion and Nature* (2005) and an MA thesis by Elizabeth Landry (1990) at the Department of Asian Languages and Literatures, University of Washington. Upon my enquiry with her supervisors (Professors Frank Conlon and Michael Shapiro), I found that neither she nor anybody else there did any further work on the Bishnois. Her own thesis is a philological study of Jambheśvara's statements based on the Hindi books by Hiralal Maheshwari. She also translated his 10 statements out of 120 statements (I have translated them all in the appendices). Her work did not deal with the environmental history of Bishnois.

other holy sites of Bishnois. Mukām is the holiest of these temples, which attracts thousands of Bishnois periodically at their founder guru's memorial shrine.

As I hired a taxi to visit Mukām at the town Nokha, I began thinking about the problems of drought and famine in this desert state that is now the largest Indian state in land area.[3] As I embarked upon my journey into the Bishnoi community, I noted vast dry land full of sand with an occasional plant or bush. I began wondering about the spread of the Thar Desert beyond the districts of Jaisalmer, Barmer, and Bikaner in Western Rajasthan. As my driver drove on the state highway, I was suddenly struck by a huge number of deer freely grazing on both sides of the highway with the entire land protected by fences and notices placed by the department of forests. I later came to know that this was one of the famous sanctuaries for blackbucks, one of the animals traditionally protected by the Bishnois. Even as the biodiversity is increasingly endangered in other parts of India and the world, the biodiversity of the desert state of Rajasthan is managed not by human isolation but by active human participation, Bishnois being one of the prime examples of it. In addition to blackbucks, other animals and birds that are found in and around Bishnoi villages are Great Indian Bustards, Indian gazelles, peacocks, bulbuls, bayas, sparrows, crows, vultures, chinkara, neelgais, wild pigs, wolves, jackals, and desert foxes. The trees that are commonly found in and around Bishnoi villages are khejari, jal, pīpal, kair, rohira, babool, and neem.[4]

With these thoughts about Bishnois and concerns for spreading desert, I reached Mukām, the place where it all began. I was soon facing what can be called the Taj Mahal of the Thar Desert. It was an elaborate white marble structure with a huge tomb in the center above the central shrine dedicated to Guru Jambheśvara. June being one of the hottest months in Rajasthan, the priest guided me to a carpeted walkway to get to the shrine. This is the most sacred site for Bishnois in the village Mukām, about 10 miles from Nokha and about 40 miles from Bikaner on the Bikaner–Jodhpur road. Like any other visitor, I was also offered the traditional hospitality by Bishnois. I was invited to lunch by the local Bishnois and was told that fresh food is kept ready throughout the year in the kitchen run by the Bishnoi community. Unlike a Hindu temple, the main shrine at Nokha does not have an idol or a statue. As mentioned above,

---

[3] Until 2000, Rajasthan was the second largest state of India after Madhya Pradesh from which Chhattisgarh was formed as a new state. Of the Indian portion of the Thar Desert, 61 percent falls in Rajasthan, 20 percent in Gujarat, and 9 percent in Punjab and Haryana combined. The greater portion lies in Rajasthan, covering about three-fifths of the total geographical area.

[4] In 1981, the village Janvi in Jalore district, Rajasthan was surveyed after 20 years and it found the similar number of flora and fauna that were counted in the census in 1961 there. This village has Bishnois and Rajputs in majority and the relationship between these two caste groups remains tense, as is the case in other Bishnoi villages historically (Padmanabha and Srivastava 1981).

Figure 4.1 The image of Jambheśvara inside a Bishnoi temple

Bishnoi temples and shrines never have any idol or statue of their guru or any other deity, even though they regard their guru to be an incarnation of Viṣṇu.[5] Only a lamp with *ghee* (clarified butter) is kept lit while praying, no elaborate ceremonies or rituals are performed by the priests. They perform *yajña*, the fire ceremony, on new moon. The temple priest claimed that the smoke created by these fire ceremonies helps cleanse the environment.[6] I asked the local Bishnois

[5] In the fifth śabda of Jambheśvara, he directs his follows not to worship any living or dead beings. Following this guideline, Bishnois worship the *nirguṇa* (formless) Viṣṇu similar to other North Indian "Sant" traditions (Gold 1987). However, I did find the pictures of Jambheśvara in the Bishnoi temples who is regarded as the tenth incarnation of Viṣṇu. In addition, they also worship several other Puranic deities.

[6] Several Indians assert the ecological aspects of performing fire ceremonies, e.g., "Some Scientific Aspects of Yagna – Environmental Effects," in *Akhanda Jyoti*, March–April 2003, bimonthly magazine of *All World Gayatri Parivar*, India.

about the lack of idol worship and a potential influence of Islam on their sect, and they outrightly rejected my question. They rather sought to connect their fire rituals with the similar Vedic practices. Similarly when I asked them about their practice of burying the dead bodies (like Muslims) instead of cremating rites (like Hindus), they argued that burying is done simply to save the firewood from the trees. One Bishnoi suggested that the practice of burying returns the body to the five elements of nature. Another potential similarity with Muslims (and Sikhs) that I found was their covering the heads before entering the shrines and temples. Although these practices are common among Bishnois today, they are not mentioned in the 29 rules laid down by Jambheśvara.

Bishnoi Community: Hindu or Muslim?

Today Bishnois are considered a caste-group within the Hindu community but in the 1891 Census of Marwar, they were classified with Muslims. Some scholars have situated them as a "liminal" community with common features from Hinduism and Islam. According to Singh and Saxena, the Bishnoi cult originated when a Rajput warrior married a Muslim girl (1998). According to Dominique-Sila Khan (2004), Jambheśvara had claimed that he was neither Hindu nor Muslim since he had raised himself above all castes and sects.[7] Traditionally, his tomb was referred to as *samādhi* by his Hindu devotees and *Mukam* by his Muslim devotees. His shrine was likewise viewed as a temple or a dargah. However, in the 1950s, after the partition of India and Pakistan that resulted in increasing polarization of Hindus and Muslims, the Bishnois were also "Hinduized" like several other "liminal" communities while a small group of Bishnoi followers joined other Muslim groups. A painting of Jambheśvara was hung on the wall above the grave and the main shrine was renamed as *mandir* with *Mukam* construed as *muktidhām*, a sacred site where *mukti* or liberation is obtained. Earlier, his grave was covered with a *chādar*, a green sheet commonly used in Islamic shrines to cover the graves of Sufi saints, but now it is covered by a saffron drapery, saffron being the

[7] The legend is that one-day Mohammed Khan Nagauri, the Muslim governor of Nagaur, who passed as guru Jambheśvara's disciple, quarreled with the king of Bikaner, also claiming to be his follower. Pointing to a number of customs, rituals, or beliefs of Bishnois, the former asserted that his spiritual teacher belonged to the Islamic tradition, while the latter insisted that he was a Hindu. Eventually, both went to see the saint. According to a Bishnoi legend, Jambheśvara replied that he was neither Hindu nor Muslim, since he had raised himself above all castes and sects (*jāt-panth*). According to Khan (1997a), Bishnoi tradition presents several features that are similar to Nizari Ismaili branch of Shi'i Islam. See Khan and Moir (2000) for similarities in the devotional hymns of Bishnois with Nizari Ismailis. According to Chandla (1998: 50), these are later "interpolations" that were introduced in Bishnoi tradition by influence from local Ismaili communities.

color associated with Hinduism.[8] Thus, an outsider visiting the shrine is easily convinced of its Hindu identity. According to Krishnalal Bishnoi, Muslims of Malerkotla town, the only town in Punjab with a Muslim majority, continue to revere Jambheśvara even today. Similarly, Bishnois also claim that Jambheśvara had inspired the Islamic sheikhs to stop cow slaughter (Rāma 2004).

However, my Bishnoi informants vehemently rejected the Muslim origins as speculated by Khan. One problem in Khan's observations is that most of her informants are Meghwals or Muslims who used to be treated as untouchables (*bhīnt*) (Gupta 1965: 4). Because of the long-standing social tensions among Bishnois and other caste groups of that region, there are several alternative narratives available about the origin and birth of Jambheśvara so the information from one group cannot be taken to be conclusive. There seem to be other inconsistencies in Khan's observations about Bishnois. For example, she mentions that Jambheśvara is a form of Krishna while he is regarded as a form of Viṣṇu by his followers. Khan asserts that Bishnois are called Prahlādpanthis because of their connection with the fifteenth-century Muslim saint Prahlāda-Tajuddin. However, one of the statements by Jambheśvara seems to contradict this: *Prahlāda su vacha kini ayo bara kaje*. This translates into a promise to Prahlāda (a well-known devotee of Lord Viṣṇu in Hindu texts) to liberate the 120 million remaining devotees. Moreover, the strong emphasis of Jambheśvara and his followers on revering the animals and trees resemble Hindu practices of pantheism and Jain emphasis on non-violence. In addition, the widespread use of Hindu symbols and myths in the statements of Jambheśvara and his criticism of Muslim practices of killing the beasts do not match the observations by Khan.

On the other hand, Khan's observations match with the conclusions of Tazim R. Kassam (1995) who has shown that the Satpanth Ismaili branch of Shi'i Islam that originated in early medieval South Asia had several Hindu myths and rituals to protect itself from Sunni persecution. Therefore, in my opinion, the syncretism present in Bishnois is a continuing tradition from the Satpanth Ismaili branch of Shi'i Islam, which had already included several Hindu myths and legends in its earlier existence before the origin of the Bishnoi tradition. This is also similar to the other syncretic movements of medieval India such as Sikhism, Kabir Panth, and Dadu, all of which show inspirations from both Hinduism and Islam in their philosophies and practices. Let us now look at this early history of the Bishnoi tradition, which was founded by Guru Jambheśvara in the fifteenth century.[9]

---

[8] None of the Bishnoi could confirm the existence of any green covering.

[9] He was a contemporary of other Bhakti saints such as Kabir (1440–1518), Surdas (1483–1563), and Nanak (1469–1538) but there is no historic record of their meeting each other. Jambheśvara resembles Nanak and Ravidas as a founder of a new community based on his new set of teachings, unlike Tulsidas and Surdas who had more "catholic" appeal (Hawley 1988). Kabir represents the Nirguna bhakti movement that spread to Dadu panth in Rajasthan and Sikh panth in Punjab (Burghart 1978). There are several similarities in the rituals and texts of the Kabir Panth and of Bishnois. Also like the other medieval

Guru Jambheśvara in the Fifteenth-Century Desert Ecology

The founder of the Bishnoi community, Guru Jambheśvara, was born in 1451 CE, in a Panwar Rajput family at Peepasar village of Nagore district in Western Rajasthan. His father Lohat and mother Hansa were devotees of Viṣṇu. Both of them had to worship Viṣṇu for a long time to seek the boon of a child, according to Bishnoi folklore. This influence from his Vaiśnava parents is obvious in many of Jambheśvara's statements. Jambheśvara regarded Viṣṇu and his incarnations very highly.[10] In turn, the Bishnoi community regards Jambheśvara as an incarnation of Viṣṇu and sees several similarities in his life with the legends of Viṣṇu's incarnations.[11] The day of his birth is the same as that of Kṛṣṇa, the eighth day of *bhādo* month of the Hindu calendar. Like Kṛṣṇa, Jambheśvara also grazed cows in his early life. His cows followed his orders just as they enjoyed the flute of Kṛṣṇa!

When he was about 25 years of age, a severe drought hit Western Rajasthan for the next 10 years. This coincides with a general period of summer monsoon failures confirmed by paleoclimatological reconstruction (Sinha et al. 2007).[12] There was a severe shortage of water and cattle fodder. People cut down a large number of trees to sell the wood in the nearby towns. The animals, such as chinkara and blackbuck, were widely hunted and killed for meat-consumption. Most of the villagers left the area in search of greener pastures. Deeply moved by this disaster, Jambheśvara began his quest for the solution. In 1484 CE when his parents passed away, he left his home and started living on Samarathal, a sand hill of Mukām village in Nokha Tehsil near Bikaner. Finally, after many years of practicing meditation and austerities, at the age of 34, he had a spiritual experience. In his vision, he saw people quarreling with nature and destroying their environment that sustained them. He decided to reform such society and concluded that humans will have to sustain the environment around them in order for nature to sustain humans. He founded the Bishnoi community on the eighth day of the black fortnight of the month of *Kārtika* of the *Vikram Samvat* 1542. His first disciple was his uncle Pulhojī. Based on his teachings about natural resources relevant in the drought years, he succeeded in conserving and protecting several local resources and soon developed a large following among the masses. Even the local ruler sought his blessings. According to Bishnoi legends,

---

saints, the biography of Jambheśvara is largely based on the hagiographical accounts of his followers and other poets in the Bishnoi community. There is very little, if any, mention of him or his followers in other historical writings pertaining to that era (Landry 1990).

10 "*Pāhal mantra*" and *Śabda-5, Śabda-vāṇī jambha sāgara*, p. 30.

11 *Smārikā*, p. 3.

12 I thank Paul Robbins (University of Arizona) and David Porinchu (Ohio State University) for this information. Chandla (1998: 83) mentions the year of this famine as 1484–5 CE. Gupta (1965: 6) mentions the year as 1486 CE.

some of the kings who met him were Sikandar Lodi of Delhi,[13] Mohammed Khan Nagauri of Nagaur, Rao Santal of Jodhpur, Rao Bikajī of Bikaner, Rawal Jait Singh of Jaisalmer, Rao Doodajī of Merta, and Rāṇā Sāngā of Mewar.

While David Lorenzen (2004) has sought to dichotomize the medieval Bhakti gurus into *Nirguṇa* and *Saguṇa* categories, Wendy Doniger has suggested that *Nirguṇa* category is a "concoction of monistic scholars" who sought a compromise between advaita and the popular love mysticism of *prema bhakti* that was directed toward the worship of images (Schomer and McLeod 1987). Based on my observations about Jambheśvara, I would agree with Doniger and other scholars who have noted the apparent artificiality of *Nirguṇa Bhakti*. Daniel Gold has dealt with some of the issues related to the tradition of *sants* in North India (1987). Jambheśvara matches several features common among *sants* of North India as noted by Gold. Jambheśvara stresses the *nirguṇa* role of Viṣṇu rather than the *saguṇa* worship as was propounded by Marathi saints such as Nāmdeva. Although Bishnoi devotion to Viṣṇu does not denote a piety in which Viṣṇu is conceived as absolutely "without qualities" – the literal meaning of *nirguṇa*, it implies that Bishnois do not perform rituals around the idols or images of Viṣṇu as some of the other Hindu communities do. On the other hand, Jambheśvara also mentioned Viṣṇu as a *saguṇa* deity, as a being with a Vaishnava name with suggestive, personifying analogies, that is different from the Sufis who talk about the divine in aniconic ways. Also like Kabir, Jambheśvara rejected the yogic practices of nāths and emphasized simple devotion using the repetition of holy name of Viṣṇu (Gold 1987). Rohatas Suthāra (1994) also notes similarities in Jambheśvara and Kabir and situates Jambheśvara in the bhakti tradition founded by Ramananda in the medieval North India. According to Hiralal Maheshwari, Jambheśvara established his sect before similar sects were founded by Nanak, Kabir, and Dadu in fifteenth-century North India. Maheshwari also situates Jambheśvara between nirguṇa and saguṇa modes of theologies. Although Jambheśvara emphasized chanting the supreme name of Viṣṇu and rejected idol-worship like other nirguṇa saints, he also acknowledged different incarnations of Viṣṇu. Moreover, the language of Jambheśvara's statements is distinct from the other saints of his time so he cannot be included with the other Bhakti saints of medieval India who all used similar language and style in their statements and verses (Landry 1990: 27). Bishnoi folksongs celebrate Jambheśvara as a combination of both nirguṇa and saguṇa (Gupta 1965: 35). I now present Jambheśvara's teachings and legends that continue to inspire his followers to protect the animals, birds, and trees in their surroundings.

---

13 For a reference about his meeting with Sikandar Lodi, Shivakumar Bhanot cites Majumdar et al. (1949: 337–41, 427–45).

Ecological Teachings and Activities of Jambheśvara

Out of his 29 rules for Bishnois, eight are about conserving and protecting the animals and trees, including non-sterilization of bulls, keeping the male goats in sanctuary, prohibition against killing of animals, cutting down of any types of green trees, and protecting all life forms. He also forbade wearing of blue clothes, because the dye for coloring them is obtained by cutting several shrubs. Seven other rules were about social behavior. They directed his followers to be simple, truthful, content, abstentious, and pure, also to avoid adultery, and making false arguments. He prohibited criticizing others and to tolerate the criticism by others. His other 10 rules were about personal hygiene and maintaining good health. These were instructions about drinking filtered water, taking a daily bath, improving sanitary conditions, prohibiting the use of opium, alcohol, tobacco, and other narcotic substances. Meat was excluded from the diet. Ritual prohibition for 30 days after childbirth and five days during menstruation was prescribed for women. Four other rules provided guidance for spiritual practices, e.g., one must always remember that God is omnipresent, perform rituals daily, and observe fast and perform communal *havan* (fire ritual) on every new moon.[14]

In addition to his 29 rules, his teachings are preserved in his 120 statements known as the *śabdas* that I have translated in the appendix.[15] I now describe some of his śabdas that relate to the environmental resources. Fifteenth-century Rajasthan had already seen the arrival and assimilation of Muslims and Jambheśvara criticized both the Hindus and the Muslims for some of their corrupt and destructive practices and habits. In his second śabda, he warned the people against harming the animals. He cautioned the people killing the animals that at their deathbed, they will hear the cries of slain animals and will have to repent for their deeds. In addition to animal protection, Jambheśvara also inspired his followers to protect the trees in many of his śabdas. In the seventh śabda, he forbade cutting trees on *Somavati Amāvasyā* (a holy new moon Monday in the Hindu calendar) and on Sundays by referring to the trees as the gatekeeper to the heavenly happiness.[16] In his tenth śabda, he criticized the people who kill animals in the name of Muhammad because Muhammad himself neither killed

---

14 Most of his teachings are still practiced by the villagers, e.g., some Bishnoi women in the villages perform *havan* (fire ritual) every morning before eating. The traditional Bishnoi villages are also some of the favorite tourist destinations. My Internet search for "bishnoi village travel tour" returned 6,830 results pointing to several travel agencies that take the tourists to Bishnoi villages in Rajasthan.

15 In addition to the translation of the complete discourse of their guru, I describe their Saṃskāras ("rites of passage") in the appendices. I also show the influence of Hindu myths and legends in the discourses of Guru Jambheśvara and the Bishnoi rituals.

16 This verse is interpreted by Bishnois figuratively. "Sunday" here is meant the time when there is sunlight (the daytime) and Somavati Amāvasyā is meant when there is

animals nor asked his followers to do so. In his tenth śabda, he pleaded for *raham*, mercy, to the followers of *Rahmān*, "if you remember that the divinity residing in your heart also resides in animals, you will surely achieve *bahiśt,* heaven." He criticized the tantric yogic practitioners sacrificing the animals to *Bhairav*, *Yogini*, or other deities and asked them to understand the real meaning of yoga. Similarly, he asked the Muslims to understand the real message of the Qurān. In his tenth śabda, he reminded the Hindus that Rāma never asked them to kill animals: "People will not be able to justify their violence when questioned by their deities." In his eleventh śabda, he forbade violence because both the action and its motive are condemnable, "The *namāz* performed by animals-slayers is hollow and meaningless. Their reading of *kalmā,* scriptures and chanting the name of *khudā,* God, will be meaningful only when they would stop killing the animals." In his sixteenth śabda, he chastised people who follow frauds as their gurus and kill animals for their rituals. He advised people to follow only those gurus who preach non-violence. In his thirty-eighth śabda, he also added that killing the creations of God in the name of God is not only wrong but is also an act of arrogance.

According to *Jambha Sāra*, another collection of his discourses, Jambheśvara advised his followers to follow six rules to avoid violence. The first rule is called the *Jhampari Pāl* that prohibits animal sacrifice. The second is called the *Jeevani Vidhi* that directs to use only filtered water and milk. The third is about putting the water-creatures back into the water to rescue their lives. The fourth is to make sure that the firewood and the cowdung for fuel do not have any creatures or insects that might be accidentally burnt. The fifth is called *badhiyā*, to avoid harming the bullocks, to sell them to butchers, or to send them to animal shelter centers. The sixth is the protection of deer in forests like cows, goats, and other non-violent beasts (Parik 2001).

Dharmic Environmentalism of Jambheśvara

The above rules laid down by Jambheśvara have been the dharmic foundation of the Bishnoi community. In his 120 śabdas, he uses the term "dharma" several times to signify both the socio-spiritual order as well as moral duty. These words of Jambheśvara are also supported by the examples of protection and conservation of environmental resources in the hagiographical accounts of Jambheśvara narrated across several generations by the Bishnoi saints and poets. Once on his way to Jaisalmer, near Nandeu village, Jambheśvara saw *tāla*, a plain piece of land, and wanted to build a water-tank there. He arranged the construction to begin there in 1510 CE that was completed in 1514 CE. This tank was known as *Jambhasar* which is about 2,000 meters wide, 400 meters long, and 25 meters

darkness, thus the cutting of trees is to be prohibited both during the daytime and during the nighttime.

deep. It is said that Jambheśvara had told his disciples that whoever will clear sand from this tank using his own labor or money, will attain heaven. This tank is near Phalodi in the district of Jodhpur. Bishnois consider this tank as sacred as pilgrimage to the river Ganga. Here, on the new moon of the Hindu month of *Caitra*, an annual fair has been held since 1648 CE. The second fair is held on the full moon of the Hindu month of *Bhādava* since 1699 CE onwards. It is believed that Jambheśvara liked this place as much as Samarathal. He stayed here for a long time and also met the famous Rajput king Rāṇā Sāngā. Jambheśvara had also built a huge water-tank at Sohajanee in district Muzaffarnagar in UP. In addition, he is credited with indicating the appropriate sites for building the new and reconstructing the old wells.

Vringali Sarovar is another such water-tank that is believed to had been built by Jambheśvara. This is located about six kilometers south-west of a village named Janglu, near Nokha in district Bikaner. During one of his visits, Jambheśvara had located a spot for pure water and later asked a local villager Varsingh Baniyal to build a tank on this spot. The tank is about 350 meters wide and 100 meters deep. On its south side is a huge mound of sand that is periodically removed from the tank in the process of deepening. The process of deepening goes on continuously for six months every year. As in the case of Jambholav, here also Bishnois ritually deepen the tank for their wishes to be fulfilled. For hundreds of years, a community called *ords* hailing from a nearby village have been employed for deepening the tank. The work was given to them on a contract basis and the money was paid by local Bishnois. The tank always supplies enough water that is freely distributed among local people. Local animals and birds take shelter in the trees around the tank. A ritual-altar is also located near the tank for performing the rituals (Bishnoi 1992).

In another such event, on *Akṣaya Tritiyā* day of the Hindu year 1572, people of Rotu village complained to Jambheśvara that their village did not have enough trees. It is believed that Jambheśvara miraculously built a huge garden of khejari trees here. Khejari trees were also planted at Lodipur, Meerpur, Mohammadpur, Devmal, and Kharad on the requests of respective inhabitants of these villages. Rotu is a Bishnoi village 60 kilometers north-east of Nagaur. This is one of the sacred places, *sāthari*, for Bishnois because Guru Jambheśvara is believed to have performed miracles here. When he visited a wedding party here, he tied the bullocks from his cart to a tree and overnight that tree turned completely green and can be found standing even today in this village. Amazed by this miracle, the villagers requested him to grow more such trees. Jambheśvara acceded to their request and miraculously grew tens of thousands of khejari trees overnight. Villagers later were concerned that the new forest would attract birds that may be detrimental to their farming. In response, Jambheśvara promised to them that the birds would only come during the night and would not harm their crops. Local villagers still believe that birds do not harm their crops and only feed on what is offered to them by the villagers. The relatively large number of trees here leads to higher rainfall than in neighboring villages. In addition, the

Figure 4.2 A Bishnoi temple near Pali

dry leaves from the trees and bird-droppings act as natural fertilizers helping achieve a greater crop-yield. Here the Bishnoi temple has preserved a sword that Jambheśvara had gifted to a local king. According to another legend, a local area remains dry throughout the year that is believed to be the path through which Jambheśvara drove away an evil person (Bishnoi 1992).

Another Bishnoi holy place, *sāthari*, is at Janglu village that has jungle of desert trees including Kāṇkeri, Kumtā, Pīpal, and Bordi. As is customary at several Bishnoi temples, here bird feed, made of pearl millets, sorghum, and moth is given to the birds, gazelles, and other animals. Water is also given to them in the earthen crucibles that are permanently placed on the ground. Both the bird feed and the water is refilled twice or thrice a day by local villagers. At the local temple here, a gown and a receptacle, used by Jambheśvara to collect alms, is preserved. While the temple was being renovated, a tree inside it was carefully preserved and the building was accordingly altered to accommodate it (Bishnoi 1992).

The next *sāthari* is at Lohavat, 109 kilometers north-west of Jodhpur. This sāthari consists of a room for performing fire rituals and four rooms where the priest and the visitors can stay. The food-grains are stored here for charitable donations and for the bird feed. Water crucibles for providing water to birds, antelopes, and other animals are permanent fixtures here like at other Bishnoi holy places. This sāthari was constructed by the Jodhpur king Maharaja Maldevjī. It is a repository for a piece of stone with an imprint of Jambheśvara's toe, according to local belief. It is believed that Jambheśvara had miraculously fed the king and his royal party here. The grateful king got a building constructed at the same spot. Animals and birds roam freely around this place and share the water from a community tank. Refilling of this water tank and feeding of birds and animals is done by local Bishnoi volunteers who periodically stay here with the priest.

Another *sāthari* is at Lalasar, about 54 kilometers south-east of Bikaner. Jambheśvara is believed to have moved here when he was 85. The place where Jambheśvara lived is on an 80-meter-high sand dune, and a Kānkeri tree stands there. There is a jungle of desert trees, including Kumtā, Borḍi, Kānkeri, and Babul, surrounding this place. Sixteen surrounding Bishnoi villages take care of this sāthari and meet here periodically for rituals. The water basins for birds and animals are also maintained by them, like in other Bishnoi places. In addition, the water is stored in the tanks by harvesting the rainwater. Jambheśvara is believed to have died here in 1536 CE (Bishnoi 1992).[17]

Examples of Dharmic Environmentalism in Bishnoi History

While media and scholars have celebrated Indian women environmentalists and activists such as Medha Patkar, Arundhati Roy, Gaura Devi, and Vandana Shiva among others (Guha 2006: 59), the story of Indian ecofeminism rarely mentions Amrita Devi who had led 363 Bishnoi men and women to sacrifice their lives to protect the khejari trees from the soldiers of the king Abhay Singh of Jodhpur.

In my fieldwork, I discovered that this widely known Bishnoi legend is challenged by the Rajput community. In June 2006, I met the current royal family head in Jodhpur, Maharaja Gajsingh, and his administrative head Mahendrasingh Nagar. I also met a number of other Rajput scholars and lay people in Jodhpur. To my surprise, many Rajputs rejected this tale of the Khejadali sacrifice. Some of the Rajput scholars encouraged me to further investigate this and bring forth the "real" fact about this legend that has now entered some of the school textbooks in Rajasthan. However, in the documentary *Willing to Sacrifice*, Gajsingh did agree about an incident that took place in Khejadali that led to the royal ordinance prohibiting hunting in the Bishnoi villages. Another local Rajput Maharaja of village Bhadrajun agreed to the Bishnoi sacrifice in a documentary made by Veg TV.[18] I visited Khejadali, the actual site of this sacrifice and found that the site has been turned into a huge martyr place adorned with a number of signboards and memorials. This site is also covered by several documentaries made by Canadian,

---

17 This was the ninth day of the dark fortnight of lunar month of the Hindu month *Mārgaśirṣa* of Vikram Samvat 1593.

18 www.youtube.com/watch?v=_1Z1knSHtQ4 (viewed on December 20, 2007).

Figure 4.3 A portrayal of massacre at Khejadali Memorial

British, German, and Indian television channels. The first major celebration of this sacrifice was held in 1978 (Weber 1990: 93).

Hiralal Maheshwari, in his seminal work on Bishnoi literature, has enumerated several incidents called the *khadānas*, where Bishnois sacrificed their lives to protect the trees (1970: 456–7). These examples are mentioned in the Bishnoi folklore by several Rajasthani poets, two by Vīlhojī, one by Kesojī, one by Nagindas, and one by Gokaljī. In addition, separate examples are also mentioned of sacrificing lives to protect animals. These are written by Vīlhojī and Sahabramjī Rahad. I will describe them briefly below.

Gokaljī (1700–90 CE) in his second *Sākhi* describes the Khejadali sacrifice. In the beginning of the verses, he praises the Jodhpur king Ajitsingh as a protector of the *dharma* because he fought against the "Turks," conquered Ajmer, and protected the forests. However, after his death, one of the ministers of Jodhpur court, Girdharidas Bhandari, plans to cut the khejari trees and 363 Bishnoi men and women sacrifice their lives to protect the khejari trees. When the then king Abhay Singh[19] hears about the incident, he stops the massacre of Bishnoi by his

[19] Although, in the traditional history written in Rajasthani language by the court historians of Jodhpur kings, this incident is not mentioned, they mention Abhay Singh as the murderer of his father Ajit Singh (Gahlot 1991: 129). There is also a mention of a

own minister and soldiers. According to Maheshwari, this extreme example of sacrifice is known by several Rajasthani words, *Sāga*, *Tāga*, and *Khadāna*. The Bishnoi poet Gokaljī advises against such extreme sacrifice. This is the only description of this incident found. Although this incident is not found in other historic texts, the people and the situations are confirmed by other texts as well. Maheshwari cites historian Ojha[20] to state that the Jodhpur ruling class faced a shortage of funds that may have forced Bhandari to order the felling of khejari trees. Minister Bhandari was a prominent political person of that time and he was appointed a minister of Merta city by Jodhpur king Ajitsingh. Bhandari is mentioned as having died in 1789 and even Jodhpur's victory over Ajmer is verified by other historic references. Krishnan Lal Bishnoi (2000) cites the handwritten manuscripts of Poet Gokaljī and the *Khyat of Khejadali Village* for this incident. In my opinion, these citations discredit the view of those Rajputs who dismiss the massacre at Khejadali as a pious mythology or a fictional account. The Rajasthan High Court also accepted it as a historic event before establishing the Amrita Devi Award.

The names and villages of the 363 people who died here are recorded by Mangilal Rao and Bhagirathrai Rao, two men from Mehlana village of Jodhpur district. Both worked for two years from 1976 to 1977 to gather this information from traditional writings. The Raos have been traditional recorders of historic events in Rajasthan since ancient times. From their research, it is revealed that people from 49 villages sacrificed their lives, 294 of them were men and 69 were women, 36 of them were married couples including one newly married couple who were passing by Khejadali village when the massacre was taking place. This event is believed to have taken place on September 9, 1730. On September 12, 1978, the corresponding day according to the lunar Hindu calendar, a large fair was held at the same village, Khejadali, for the first time to commemorate the massacre.[21] The fair has now become an annual event organized by All India Jeev Raksha Bishnoi Sabha, a national Bishnoi NGO for wildlife protection.

---

building project that remained incomplete (Nainsi and BhātÚi 1968: 567), possibly because of lack of fuel due to the resistance of Bishnois in giving the khejari trees demanded by Abhay Singh's officers. There is also a mention of political influence of Bhandaris, the caste of officer who ordered the felling of khejari trees and the massacre of Bishnois (Gahlot 1991: 131).

20 Ojha, *Jodhpur Rājya Kā Itihāsa* 1941.

21 The fair was reported in Hindi newspaper *Navbharat Times* of September 3, 1987. Coincidentally, these were the years when Chipko in UP, Appiko in Karnataka, and Anna Hazare's efforts in Maharashtra were being hailed as successful socio-ecological movements based on people's participation. Swadhyaya also built their first tree temple in the same year. Apparently, Bishnois started realizing the importance of their ecological traditions after the emergence of other socio-ecological movements, although their other fairs at Mukām and Jambholav have been celebrated from much earlier times.

Figure 4.4 The Sacrifice Memorial at Khejadali

R.J. Fisher argues that Bishnois have also used their religious and ecological beliefs as political symbols (1997: 64–74). Mentioning the famous tale of sacrifice of 363 Bishnois against the soldiers of the Jodhpur king, Fisher notes that the Bishnoi rules were used in two ways. First, for the Bishnois, the khejari tree symbolized the political resistance to the Rajputs. Second, the king of Jodhpur decided that honoring the beliefs of the Bishnois would be a better way of establishing control than by attempting to override them. However, seeing the above sacrifice legend in the context of similar contemporary examples shows that the Khejadali sacrifice is not just a past memory for Bishnois; it is still an active practice. While the Khejadali sacrifice has recently received some attention, I discovered several other similar incidents from Bishnoi manuscripts and from Indian newspapers. I now describe all such examples that I discovered in my research.

In 2001, the Indian union environment and forest ministry awarded the first Amrita Devi Award to Gangaram Bishnoi posthumously (*The Hindu*, May 30, 2003), who sacrificed his life trying to protect a chinkara deer on August 12, 2000 (*The*

Figure 4.5 Bishnoi farmers with khejari trees in their farm

Times of India, August 15, 2000). Gangaram Bishnoi, 35 years of age, was working in his fields on August 12, 2000. At about 5pm, he saw a person, Peparam, with a gun, taking aim at a wild chinkara grazing nearby. Gangaram intervened to save the antelope from poachers but Peparam fired and the chinkara fell on the spot. Peparam and his accomplices lifted the chinkara and started running away. Gangaram chased the poachers along with his brother, ran after them for approximately three kilometers, and caught the poachers. In the ensuing struggle, poachers fired at Gangaram who died on the spot. He was survived by his wife, two sons, and three daughters, who expressed their pride that Gangaram laid down his life following the Bishnoi tradition of protecting the wild animals.[22] Once again, in 2004, this award was given to a Bishnoi, Chhailuram Singh Rajput of Bhiyasar village in Bikaner district, who sacrificed his life to save blackbucks.

In another well-known incident, the Hindi film actor Salman Khan was sentenced to five years' imprisonment for killing a blackbuck,[23] the sacred gazelle

[22] http://pib.nic.in/archieve/lreleng/lyr2001/raug2001/24082001/r2408200117.html.

[23] This was widely reported in the media (for instance, see *The Frontline*, April 22, 2006 and the *New York Times*, November 29, 1998). Even a feature film, *Qaidi Number 210*, was announced which was supposed to be made with real-life characters including Mahipal Bishnoi, who fought the case as a representative of the Bishnoi community,

of the Bishnois, on September 26, 1998. This has been possible largely due to the active involvement of Bishnois in the entire legal process. In January 2007, local Bishnois of the village Agneyu in Bikaner filed complaints against another film producer when a horse died on set. On March 14, 2008, the Akhil Bhartiya Jeev Raksha Bishnoi Sabha demanded the removal of Indian cricketer Mahendra Singh Dhoni from the national cricket team for sacrificing an animal (*New India Press*, March 14, 2008). In October 1999, Bishnois surrounded the local police station in Churu, Rajasthan, after more than 20 Indian gazelles and three peacocks were found dead near the village of Sansatwar. Authorities had to suspend the local police officers for their alleged negligence in failing to prevent these killings (BBC News, October 28, 1999). Another episode of Bishnoi environmentalism comes from the Abohar Wildlife Sanctuary. Its Divisional Forest Officer regularly depends on the local Bishnoi community in night patrolling against the poachers (*The Times of India*, June 8, 2003). In Haryana also, Bishnois are often the first to report poaching incidents (*The Times of India*, January 12, 2003).

There are several similar incidences of Bishnoi activism throughout their history. Consider the following events mentioned in the Bishnoi tales and legends. Vīlhojī was the fourth successor of Jambheśvara and lived from 1532 to 1616. In his seventh verse, *Tilavāsani Sākhi*, he describes a similar incident. Two women, Kheevani and Netu Naina, and one man, Motojī, sacrifice their lives to protect khejari trees from the attack of Gopaldas Bhati. This took place in Tilavasani village of district Jodhpur. Sahabramjī Rahad (1814–91) has also described this incident in his book *Jambha Sāra*. Like the Khejadali massacre, this incident also transformed the killer officer and made him a kind man witnessing the deaths of innocent Bishnois. Similarly, in his ninth verse, two women, Karma and Gaura, sacrifice their lives to protect khejari at Revasadi village in Jodhpur district from the local property owners in 1604. Poet Nagindas describes a similar sacrifice by Ramdas of Nagaur. Poet Keshodāsjī Godara (1573–1679) describes Buchojī Echara's sacrifice at Polawasa village of Merta against Narsinghdas Thakur of Rajod in 1643. Similarly, poet Sahabramjī Rahad, in the twenty-fourth chapter of *Jambha Sāra*, describes a similar sacrifice to protect bulls in Cheengarha village in Hisar district of Haryana state in 1857. This also describes a protest by Tarojī against British hunting of deer in Seeswal village of Hisar and royal hunting in Heengoli by Jodhpur king Takhatsingh. Takhatsingh later issued a copper plate with an order to ban hunting in Bishnoi villages.

Another sacrifice was made by Gorkharam and his sons Pratapram and Chimnaram of village Barasan in district Barmer in April 1947. They were bringing water on their camels when they saw a herd of antelopes being chased by poachers. Both the brothers chased the poachers. However, as is often the case, unarmed Bishnois were killed by the poachers with deadly weapons. The

---

Hastimal Sarawat, Salman Khan's advocate, and the driver of the car used in the killing of the animals (*Down To Earth*, September 30, 2007).

spot where both were killed was in a farm belonging to a farmer who has left that piece of land untilled as a tribute to their sacrifice.

In February 1948, Arjun Ramjī Bishnoi of village Bhaktasani in district Jodhpur received bullet injuries in his stomach when he protested against poaching around the village pond. He succumbed to his injuries the next day in Jodhpur Hospital. Chunaramjī Bishnoi, son of Hardonjī Godara, resident of village Rohicha Kallan, district Jodhpur, was killed in 1948 while trying to protect a herd of blackbuck and chinkara from illegal hunters. Yet another Bishnoi, Bhinya Ramjī, of village Banad in district Jodhpur, was killed in May 1963 on the back of his camel while trying to intercept poachers. Similarly, Dhookalram Bishnoi caught hold of the loaded gun of a poacher named Mangalasar, as the latter was aiming at a chinkara. The gun was fired at him and he later died on his way to the hospital.

Birbal Ramjī Bishnoi of village Lohavat, district Jodhpur, was killed in December 1977, while protecting a gazelle from a group of three poachers. One of the killers was sentenced to rigorous imprisonment for 20 years by the Sessions Court, the second was imprisoned for one year, and the third was acquitted. The village Lohavat had not contributed to the list of 363 martyrs of Khejadali but, with this 1977 sacrifice, it has also joined the distinguished Bishnoi villages with examples of "martyrs" to save animals or trees. Dozens of similar sacrifices are noted by Hanumana Singh Bishnoi in his Bishnoi journal *Sangoṣthī Vāṇī* (Brockmann and Pichler 2004).

In 1995, Bishnois caught an Indian Air Force captain hunting. In October 1996, Nihal Chand Bishnoi was killed while protecting a deer from a gang of poachers. His father Hanuman Singh Bishnoi was quoted in the documentary *Willing to Sacrifice*, "My son was killed by poachers, when he tried to save a deer from them. I grieve for him, but I am happy, he became a martyr, while protecting the Bishnoi faith." This documentary, produced by Yamini Films and directed by B.V.P. Rao, won the award for the Best Environment Film at the fifth International Festival of Films, TV and Video Programme held in Bratislava, Slovakia. In October 1999, India's President K.R. Narayana awarded *Shaurya Chakra* to Nihal Chand Bishnoi posthumously. A similar incident took place in the late 1970s, in village Mehrana near Abohar in district Ferozepur, Punjab. A young girl, Sharda, rushed to protect a wounded blackbuck as soon as she heard a gunshot and even attacked one of the poachers. Later, one of the poachers was produced before the village and was handed over to the police. For her heroic deed, she was honored at the annual mass fair at Mukām in district Bikaner in Rajasthan.

Another incident to show Bishnois' love for animals took place in village Nadori of district Hisar in Haryana on May 10, 1978 when Mrs. Rami Devi, wife of Rameshwar Das Dharnia Bishnoi, breast-fed a fawn to save his life.[24] The incident took place when a party of hunters chased a deer herd. A pregnant doe straggled

---

24 www.youtube.com/watch?v=rmAWi8uSak0 (a small video of a Bishnoi woman feeding her baby and a fawn, viewed on December 20, 2007). *Hindustman Times* lensman

behind under cover of bushes and delivered the fawn.[25] The young fawn was nurtured by her until he grew old enough to be delivered to the local zoological garden. Also in 1978, Bishnois prevented Arab sheikhs from hunting the Great Indian Bustard in Jaisalmer. The Indian government also took timely action and this bird was saved from total extinction. This bird is now the official state bird of Rajasthan. It is also a favorite bird of Bishnoi farmers because it consumes locusts, rats, and other pests harmful to farming.

Apparently, even animals and birds have started recognizing protectors from invaders. Since Bishnois wear traditional clothes with familiar colors and patterns, animals feel comfortable in their presence while they run away from outsiders. Even Bishnois themselves get suspicious when outsiders enter their villages having witnessed several examples of poaching and hunting. A noted ecologist, Kailash Sankhla, described an incident when he visited a Bishnoi village with Peter Jackson in the 1980s. At first Bishnois were suspicious of both of them but later welcomed them in their customary ways when they came to know about the intent of their visit (Sankhla and Jackson 1985).

In the 1940s, before India's partition, a train carrying British soldiers was passing through a Bishnoi village in the Muslim princely state of Bahawalpur (now in Pakistan). When a British soldier fired at the animals in the Bishnoi village, Bishnois attacked the train and a complaint was filed against Bishnois that was later dismissed. In another incident also at the same village, one of the maternal uncles of the Nawab of Bahawalpur wanted to hunt in the Bishnoi village. When he attempted to hunt, Bishnois protested vehemently and ultimately the Nawab acceded to the Bishnois and banned the hunt there.

In 1975, Bishnois established a nationwide wildlife protection organization called All India Jeev Raksha Bishnoi Sabha. This is a Bishnoi organization for the protection of all forms of life and its membership is open to all, including non-Bishnois, who are interested in promoting animal rights. The organization was initially founded in 1966 as the "anti-hunting committee" and later was changed to the present name to make it sound more positive rather than negative. On January 15, 1975, it was registered in Punjab state. As noted above, this organization was the main reason for the banning of the hunting of the Great Indian Bustard in Rajasthan in 1978. In December 2000, India's noted environmentalist and political leader Maneka Gandhi paid tribute to its co-founder Sant Kumar Bishnoi and unveiled a statue in his honor in Abohar, Punjab. He was also the winner of the Indira Gandhi Paryāvaraṇa Award, a national environmental award. He had built a formidable reputation as a savior of animals and an environmentalist of national repute. His operations in the wildlife sanctuary, spread over 16 villages of the subdivision, were primarily responsible for the protection of blackbucks,

---

Himanshu Vyas won the IFRA Gold Award for his picture of a Bishnoi woman breast-feeding an orphan fawn (*Hindustan Times*, April 4, 2008).

[25] *Beyond the Tiger*, written by M.K. Ranjit Singh, shows a photograph of Rami Devi suckling a blackbuck fawn from her breast.

an endangered species. He donated his entire award money to environment protection causes (*The Tribune*, December 10, 2000).

In another incident, on April 8, 1978, three hunters of a town Fatehabad in Haryana went to Muthan village and fired at a doe. The animal was injured and escaped to the fields of Keharsingh. Keharsingh, trying to save the injured animal from the hunters, was killed by the hunters. Even though he was not a Bishnoi, All India Jeev Raksha Bishnoi Sabha organized a public meeting on June 20, 1978 on the Grand Trunk Road where he had sacrificed his life. In addition to raising public awareness about his sacrifice, the Sabha also helped the bereaved family financially. The Bishnoi youth magazine *Sanghoshthi Vāni* issued a special edition to honor the martyr.

Another case that the Sabha pursued also involved a non-Bishnoi. Harinarayana Bajpai burnt himself alive on May 11, 1983 on the bank of river Sangur, near the forest in Kanpur district, Uttar Pradesh, where a blue bull (*Neel Gāya*) was being exterminated by poachers. This 23-year-old man had given a 15-day notice before he took this extreme step. When the Sabha heard about him on May 15, 1983 in Punjab through a news report, Bishnois immediately came into action and sent the reports to various government officials in New Delhi and Uttar Pradesh. When they did not get any response from the government officials, they announced that everyday one Bishnoi would follow the footsteps of Harinarayana at the same site where he had immolated himself. Fortunately, on August 13, 1983, Uttar Pradesh government banned all kinds of hunting. The Sabha later built a memorial in the honor of Harinarayana near his village. It also honors noted ecologists at the annual Bishnoi fairs. It also extended cooperation in 1988 to the Indian branch of the international NGO Beauty Without Cruelty in the latter's task of banning the merciless killing of karakul lambs in government institutes, for the sake of pelt. The Bishnoi members of the Sabha in Abohar area are active in protecting the animals and birds in the Abohar sanctuary in which thousands of blackbucks, peacocks, hares, and other animals and birds are conserved without adequate measures. The neighboring Bishnoi villages protect this open sanctuary without any barbed wire or any other fencing.

During my trip, I also met Gurvindar Bishnoi in Jodhpur who had founded an NGO Community for Wildlife and Rural Development Society. Like other Bishnoi examples, his mission is to save and protect animals that are injured by accidents or by hunters (*The Times of India*, April 11, 2006). Whenever a deer or blackbuck or any other animal or bird is injured, people call Gurvindar Bishnoi for help. He rushes to the location, takes the injured animal to the hospital, and takes other legal action if necessary against the hunter. He had also produced a video documentary about Bishnois and Jambheśvara.

I also visited Śri Jagatguru Jambheśvara Goshalā Sansthā, at Mukām. This cow shelter institution takes care of more than a thousand cows. Their food is brought from Haryana and Punjab. About 8,000 to 9,000 kilograms of food is fed to them on a daily basis. This institution is inspired by Amar Thāt, an animal shelter institution mentioned in one of the verses by Jambheśvara's disciple

Udojī Naina.[26] Jambheśvara had prohibited keeping goats as pets and had ordained against the slaughter of goats and sheep in another verse.[27] Currently, there is only one Amar Thāt in village Rotu, in district Nagaur in Rajasthan. As a rule, following the Holi festival, villagers participate in a public auction to take care of the Amar Thāt for the next year.

Bishnois also successfully influenced several ruling classes to declare prohibition against hunting and poaching. In 1752, Maharaja Anupsingh of Bikaner issued a notification to ban the cutting of green trees. In 1878, Maharaja Mansingh of Jodhpur issued a notification to ban the cutting of khejari trees. Also in 1900, Takhatsingh, king of Jodhpur, banned hunting in Bishnoi villages. Even the British Raj issued similar orders in Hisar and Ferozepur districts of Punjab in 1896 and 1898. Similarly, in 1901, the Bikaner king issued another such declaration.

P. Sivaram, a sociologist at the National Institute of Rural Development, Hyderabad, conducted a study at two Bishnoi villages in Luni block of Jodhpur district in 2000.[28] The respondents mentioned that they were staunch followers of 29 principles due to which their cattle population, green patches, and soil fertility have increased. Based on these benefits, Bishnois were more prosperous than other local communities were. He also found several sacred groves in the villages managed by Bishnois, including some that were claimed to be about 400 years old.

Based on several examples of environmentalism practiced by Bishnois, it is evident that the religious charisma of their founder guru has successfully endured over several centuries.[29] Bishnois act as a deterrent against the hunting expeditions of outsiders. The animals and birds tend to concentrate near Bishnoi houses during the late afternoon and early evening, which are the common

---

26 *Bakrā pāley thāt kar, tanni nahin nakho*, which means that the goats should be looked after in thāts and bullocks should not be castrated.

27 *Kinnri tharpi chhali roso kinnri Gadar gai, sool chubhijey karak duheli to hai jayo jeeva no ghai*, which means by whose sanction do butchers kill sheep and goats? Since even a prick by a thorn is extremely painful to human beings, is it proper to indulge in those killings? Therefore, these animals should be treated as own kith and kin and should not be harmed in any way. In another verse, while preaching to another disciple Nathajī, Jambheśvara says, *Chhery bheri ādi ko par upkāri mann, rakshā main tatpar rahey so buddhimān*, which means that goats, sheep, etc. are rendering service to others, and the one who protects them is a wise person.

28 http://nird.ap.nic.in/clic/reshigh_2k_13.html (viewed on January 10, 2007).

29 However, one violation of the 29 rules is their consumption of opium. Bishnois have a long tradition of opium usage. Opium tea was used as a welcome drink for guests. Other traditional occasions for opium tea are when the younger members of the family serve it to the elders of the family coming back from work and gatherings such as marriages and funerals. Other such violations include smoking, use of alcohol, wearing blue clothes, eating outside, skipping the fast on new moon, and not taking a bath in the morning.

times for hunting. There are also specific areas reserved in their villages where animals and trees are "religiously" protected.

Bishnoi Dharmic Sanctuaries for the Plants and Animals

The Bishnois manage sacred groves called orans in the arid and desert regions of Rajasthan. Despite sparse vegetation and limited water resources, the area supports a higher density of human and animal populations than any other desert region in the world because of the conservation practices of its people (Islam and Rahmani 2002; Gaur and Gaur 2004). According to Paul Robbins, the Thar is frequently referred to as the most densely populated desert in the world. While it is difficult to categorically verify this claim, the desert districts of Rajasthan are indeed densely populated, including large urban and rural human populations in a highly water and resource scarce environment. According to the 2000 Census of India, the desert districts of Rajasthan, Jaisalmer, Barmer, and Jodhpur had nine, 51, and 94 people per square kilometer respectively (Government of India Directorate of Economics and Statistics 2003). In comparison, the population density of Egypt is approximately 66 persons per square kilometer and that of Saudi Arabia approximately 11.

The basic philosophy of the Bishnoi faith is that all living beings have a right to live and share resources so humans should try to minimize harms done to other creatures as much as possible. They follow the rules laid by their guru including a ban on killing animals and on felling trees, especially their most sacred khejari tree that has numerous life-sustaining properties. The orans also provide a protective habitat for the Indian gazelle and blackbuck (*The Tribune*, June 18, 2005). There are only a handful of documented sacred groves in Rajasthan, but their areas often exceed thousands of acres. According to a study published in *Down to Earth* magazine in 2003, Bishnois maintain nine sacred groves in a 241-hectare area (Sethi and Viswanath 2003).

Jha and his team conducted a study for the World Wildlife Fund on the status of orans in Peepasar and Khejadali villages of Rajasthan where Bishnois are in majority. They found that some of the orans were about 200 years old. The birthplace of Jambheśvara, Peepasar, has five sacred groves, *Jambheśvara kī oran* (31 acres), *Pīpenjī kī oran* (11 acres), *Maharaja kī oran* (42 acres), *Hanumānjī kī oran* (five acres), and *Oodjī kī oran* (two acres) thus covering an area of 90 acres. In 1916 CE, a fire disaster racked the small village of Peepasar. After one year, villagers dedicated several pieces of their personal land holdings for orans where only *gochar*, grazing by cows, was allowed. At Jambheśvara kī oran, there are 190 trees of khejari, Kumta, Kankeri, Rohida, and Bordi and other desert species of bushes. The animals here are the gazelle, blackbuck, fox, and cat. Birds present include the kestrel, vulture, peafowl, lapwing, owlet, bee-eater, bulbul, roller, babbler, robin, wagtail, and tailor bird. At Maharaja kī oran, there are 156 trees of khejari, Kankeri, Bordi, and Rohida. The animals are the same as at Jambheśvara kī oran.

Adjacent to the oran is a village pond that acts as a natural reservoir of the rainwater and is used by villagers for drinking water. The caretaker of this oran, Maharaja, leases out its land for agriculture and some portion of the money is used to maintain the local temple. When not being used for agriculture, the oran also serves as pasture for livestock of the nearby village. Pīpenjī kī oran has nine khejari trees. It is linked to the local deity *Pīpenjī Bhagat*, a Rajput devotee of Śiva in whose name village Peepasar was named before the birth of Jambheśvara. Hanumānjī kī oran is linked to Hanumānjī and has only five khejari trees (Jha et al. 1998).

Now I describe some of the benefits of khejari[30] trees that are based on the accounts of R.S. Bishnoi (1992), Mann and Saxena (1980), and farmers whom I personally interviewed. The khejari tree is one of the rare botanical species that grows in deserts because it requires little groundwater for its survival. Its leaves make excellent fodder for camel and cattle. Its crown produces a legume known as *Sāngri* that is rich in protein and is used in Rajasthani food. Its trunk is used as timber for making agricultural implements and for building purposes. Droppings from the birds using it as their shelter constitute valuable manure for the farms where these trees are located. In addition to providing shade, these trees also keep the temperature low in the high temperature zone of Rajasthani desert. Timber and firewood from these trees are taken only from their dead branches. Its dropped leaves can be used for fodder, its dropped branches for fuel, and its fruit for food. Khejari trees stabilize sand dunes and they are said to increase yields of crops that grow close by. Another tree mentioned in the Bishnoi literature is *Kankeri*. Guru Jambheśvara mentioned that in his meditation he sat near this tree. This is a small plant of about two to three meters of height with a dark green crown. This dense plant serves as an ideal cool shelter in the hot desert for animals and birds. Its fruits attract a large number of insects.

Blackbucks that are protected by Bishnois against the poachers are included in schedule 1 of the Indian Wildlife (Protection) Act, 1972, as endangered species. Today, the population in India is about 48,000. About 90 percent of it is in the Bishnoi areas of Rajasthan, Punjab, Haryana, and Gujarat. Like khejari trees, blackbucks also survive in the desert based on their unique characteristics. Unlike humans and other animals, blackbucks can drink water with high salt content. This enables them to survive even where water is not of good quality. Their main food is leaves from the desert plants such as khejari. However, they also ingest a lot of soil that transforms into useful fertilizer with their digestion process. In Haryana, their number was 4,193 in 1993, 75 percent of which was in the districts of Hisar, Fatehabad, Sirsa, and Bhiwani. Punjab has only one sanctuary covering an area of 13 villages in Abohar tehsil in Ferozpur district, adjoining Shriganganagar district in Rajasthan. There were about 4,000 blackbucks in 1993, up from 3,500 in 1988, as reported by S.K. Sharma, Environment Society

30 Khejari is mentioned as *Śami* in several Sanskrit texts, e.g., MSm, RV, AV, ŚBr, Ramayana, and MBh (Mann and Saxena 1980).

Chandigarh (*The Tribune*, October 19, 1998). The Bombay Natural History Society co-directors, M. Zafar-ul Islam and Asad R. Rahmani, who carried out extensive research on the blackbucks, found that their maximum number is found at Guda-Bishnoi and Dhava-Duli villages of Jodhpur district, Rajasthan (Islam and Rahmani 2002). Dhava-Duli is also declared as a wildlife conservation park. The highest density is found in Tal Chhapar sanctuary in Churu district that has 2,000 blackbucks (*The Times of India*, October 17, 1998). All these towns and villages are largely inhabited by Bishnois.[31]

Concluding Remarks

Although deep ecologists also believe in preserving nature for its intrinsic value, they see the human intervention with nature as harmful. They seek to preserve forests and other areas, free from humans (Sale 1986). However, communities such as Bishnois have maintained a fine balance with nature for several centuries. Like Bishnois, other farming communities and native hunter-gatherers used natural resources but maintained the balance. Scholars such as Ramachandra Guha (1999) have argued that it is neither practical nor advisable to remove all human intervention from nature as proposed by deep ecologists, especially in Asia and Africa. Human intervention has often sustained the ecology. The Indian government has already started incorporating human participation in its forest management starting from 1988 (Jeffery 1998). While the ruling Forest Policy of 1952 had stressed state control and industrial exploitation, the new Indian National Forest Policy of 1988 emphasized the imperatives of ecological stability and people's needs (Guha 2006: 119). Going even further in 2006, the Indian government passed a new bill seeking to recognize and vest the forest rights and occupation in forestland of forest-dwelling Scheduled Tribes and other traditional forest-dwellers (*The Hindu*, December 16, 2006). This agrarian environmentalism is different from both the progressive conservatism, which places human needs above nature, and deep ecology, which subordinates

---

[31] In contrast, in Pakistan, blackbucks are near extinction because of excessive poaching, predation, habitat destruction, diseases, and inbreeding. According to a news report (*The Hindu*, February 24, 2008), Pakistan had to import blackbucks from Texas, USA during the late 1970s and early 1980s, for the purpose of a "reintroduction programme." However, these projects have not been fruitful. Today, blackbucks exist in small numbers in three or four locations in Pakistan, particularly in Kirthar, Sindh, and Lal Sohanra National Parks in Punjab. Over 413 animals are being currently bred in the Lal Sohanra Park. More blackbucks are provided to the conservationists from time to time for further conservation in Lal Sohanra's private reserve. At present, about 500 blackbucks are surviving in the Mir of Khairpur Mehrano reserve, 110 at Khangarh, and 70 at New Jatoi, Nawab Shah. Apart from these, a small number of them are kept in zoos, wildlife centers and private farms and houses (*Daily Times*, July 28, 2004).

humans to nature (Parajuli 2001). Alternatively, as Ann Gold writes about the village ecologies of Rajasthan (2001), "Converging individual, communal, geophysical, and cosmic interdependencies are both systematically ordered and perpetually negotiated in these locally posed ecologies." We can identify two issues here. One is the issue of human intervention that can be positive such as in the case of the Bishnois. Alternatively, it can also be negative as in the case of ecologically unsound development projects, e.g., the Tehri Dam. The other issue is the intrinsic as opposed to extrinsic or instrumental value of nature. Evidently, for the Bishnois, nature is both intrinsically valuable for its own sake and extrinsically valuable for all the benefits that this desert community derives from it.

Environmentalism is important in the third-world countries for their sheer survival, as exemplified by Chipko movement that is different from the environmentalism of Western organizations, such as Earth First, meant for enhancing the quality of life. Deep ecology also tries to link itself with Eastern and other native religious traditions to portray its universal appeal. However, it ignores that Eastern traditions themselves have also remained anthropocentric rather than biocentric. Humans are accorded a special status in all Indic traditions, rather than rendered just a creature as done by deep ecologists. Eastern societies have not shied away from using natural resources for their needs; they only did it in a sustainable way. The friendly climate and inherent paganism may have helped in this regard. Overall, to term Eastern traditions as a foundation for deep ecology is ignoring several facts and so is not accurate. On the other hand, Ramachandra Guha in his enthusiastic critique of Western approach of "deep ecology" has ignored dharmic communities and movements that have preserved and sustained natural resources beyond their limited economic needs. Swadhyayis, Bishnois, and Bhils (as I will show ahead) have practiced their environmentalism that cannot be categorized under Social Ecology as Guha did to the Chipko movement (Guha 2006). My case studies would rather fall under Dharmic Ecology. I can only partially agree with O.P. Dwivedi who categorized Chipko as an example of Dharmic Ecology (2007) because dharmic symbols were incorporated into the Chipko movement only in its later phase. It was primarily a movement with a long history of protest against the authorities as shown by Ramachandra Guha (2000) and Arun Agrawal (2005). In contrast, the environmentalism of Swadhyayis, Bishnois, and Bhils is primarily a reflection of their dharmic traditions. Their environmentalism has very little, if any, motivation based on their economic survival that was a major force for the activists of Chipko movement.

Van Buitenen underscored the importance of vernacular texts for understanding Hinduism (cited by Peter van der Veer 1989), "Anthropologists, who wished to collaborate with textual scholars in their endeavor to understand modern Hinduism, should take the study of much more recent (vernacular) texts as their starting point." In the appendix, I have provided the complete translation of Jambheśvara's śabdas to show that Jambheśvara's statements not only show their connections with the so-called "Brahminical Hinduism" but also show the connection of "dharma" with ecology. Most of his statements bear the strong influence of Hindu cosmology and Hindu myths, including the Rāmāyaṇa, the MBh, and the Purāṇas. This is different from Emma Tomalin's assertion that ecotheologians only offer *interpretation of traditions*, a reconstruction, and reinterpretation of traditions to connect them with environmentalism (2009). Like the case of Athavale, Jambheśvara also offers traditional interpretations of Hindu myths and legends without any major reinterpretation or reconstruction. He simply reinforces the powerful influence of the term "dharma" among his followers. Introducing his autobiography, Mahatma Gandhi had mentioned, "In my experiments, spirituality means morality, and dharma means ethics, morality practiced with a spiritual view is dharma." Like Gandhi, Jambheśvara also saw morality, ethics, and spirituality intertwined. For example, in his seventy-second śabda, he emphasizes that only moral virtues such as truthfulness, honesty, and compassion evict the evil from a human and transform oneself into a true human being. In his eleventh, twenty-third, seventy-seventh, ninety-ninth, one-hundred-and-sixth and several other śabdas he emphasizes moral actions as a prerequisite for religious life.

Overall, we see the same overlap of religious, personal, and ecological attitudes that we see in the Swadhyaya movement's ideologies laid down by Athavale. Again, like the Swadhyaya movement, the term "dharma" is used interchangeably both by the founder and the followers of the Bishnoi community. However, what is different in the Bishnoi case is their historical examples and legends that their poets and hagiographers have preserved for several centuries. These examples continue to inspire more Bishnois to transcend their religious ethos and fight to protect and conserve the trees and animals even at the cost of their personal lives. Most of them are barely aware of the Western scientific discourse about "global warming" or "biodiversity." For them, their tradition based on the words and life of their guru is sufficient to take up the cause of environmentalism. This is much beyond the recognition of bio-divinity based on the Hindu cosmology or Hindu texts. We see more clear evolution of the textual or ritualistic reverence for trees and animals into practical everyday implementation of ecological activism. Thus, Bishnois serve as one of the most powerful examples of environmentalism that is rooted in their dharmic tradition. Unlike other Hindu communities, castes, and movements, the dharma of Bishnois is not just limited to their religious rituals or scriptures, but it includes natural resources much beyond their religious sites as is evident from the examples of their sacrifices done in the farmlands of their villages.

Chapter 5
Sacred Groves of Bhils

How can we cut the trees of the God?[1]

India today has approximately 13,720 sacred groves in at least 19 states. Kerala, Maharashtra, Andhra Pradesh, and Tamil Nadu have the maximum number of sacred groves (Malhotra et al. 2007). These groves are rich in rare and endemic species of plants and represent a tradition of conservation, management, and even sustainable development of natural resources (Ramakrishnan et al. 1998). The finest such forest of India appears to be in the Sarguja district of Chhatisgarh. Here every village has a grove of about 20 hectares.[2] These groves, known as *Sarana* forest, traditionally have served as sanctuaries for animals. In Kerala, there are about 750 such groves, of almost one square mile, whereas orans constitute 6 percent of the land area in Uttara Kannada. In Gujarat, orans are among the highest rank in 12 categories of Community Conservation Areas (CCA) (Singh n.d.). Recently, a debate has arisen over whether these groves can play a role in environmental awareness. In this chapter, I first survey the existing literature and then present evidences from my fieldwork in Rajasthan that show the ecological role of sacred groves contrary to the intense skepticism of some scholars.

In her book on sacred trees, Albertina Nugteren (2005) provides several examples of sacred groves that show the interaction of material and symbolic values. She concludes that Indian cultural embeddedness in the natural world, exemplified by sacred groves, can greatly enrich and enhance the current ecological movements. She also cautions that the claims about the intrinsically green nature of Indic religious traditions can be exaggerated and can actually harm the effectiveness of environmentalism as a social movement.

Rich Freeman argues that sacred groves in Kerala are not based on environmental awareness. He concludes that these groves are not maintained out of an attachment to nature or a desire to protect it. According to him, the products of such groves are not consumed by the local people out of religious sentiments for the deity associated with them (1999: 257–302). Instead of rejecting the ecological importance of sacred groves, Vasan and Kumar present

---

1 This is my translation of the words of the people at a Bhil temple in Banswara district, Rajasthan. This chapter is largely based on my fieldwork in that area in 2006. All the photos are taken by me unless otherwise indicated. I thank Elizabeth Kent for her helpful comments on an earlier draft of this chapter.

2 One hectare equals 10,000 square meters or approximately 2.5 acres.

two contrasting examples of sacred groves, *devbans*, from Uttaranchal (2006). One of them is a model of community conservation based on an indigenous belief system and management practices, but the other suffered the opposite fate. While the latter is a challenge to simple models that view sacred groves as archetypical institutions of community conservation, the presence of the former shows that straightforward critiques of community-based conservation present only a partial picture. These two contrasting examples demonstrate that Freeman's critique of Kerala's sacred groves present only one side of the story. Arun Agrawal's observations are similar from his fieldwork in about 40 villages in the Kumaon area of Uttaranchal (2005). He notes the transformation of villagers, who set hundreds of forest fires in the early 1920s to protest British state regulations, into conservationists by the 1990s. It is the "environmentality," the interplay of governmentality and environmentalism, that explains the transformation of those villagers. Obviously, any one-sided explanation of environmentalism – or lack of it – cannot do justice to this complex phenomenon.

Another argument made by skeptics is that the physical inaccessibility, and not the religious sentiment, prevents the exploitation of natural resources in the sacred groves. Kalam's extensive study of Karnataka's sacred groves, called *devarakadus*, proposes that there are two kinds, "those where encroachments ... have taken place, and those where there is almost no encroachment." Kalam imputes a straightforward materialistic causality to this; the former are "close to human habitation" and the latter are "far from human settlements." Thus, Kalam concludes, it is only around these more isolated groves that "aspects of awe, mystery, wrath of the deities, gods, or goddesses, legends of punishments, etc. are woven, and survive to this day" (1998: 42–3). However, Ann Gold presents quite a different conclusion in her research on the sacred grove of the Malajī Temple in Bhilwara district, Rajasthan. She found that the sacred groves at this temple are protected by the combined effect of the ongoing faith of Mautis Mina community in Malajī's divine order and a thoroughly systematic and rationalized model of landscape protection (2006).

Following Ann Gold, I now present examples from the Bhils who have protected and conserved the sacred groves in the districts of Banswara and Dungarpur in Southern Rajasthan. Some of these examples show an active participation of both the community and the governmental initiatives while other groves continue to survive based on indigenous religious beliefs alone. Before presenting my examples, let me present a few examples of similar studies done in the other parts of Rajasthan.

Sacred Groves in Rajasthan: Some Examples of Recent Studies

Sacred groves are known under various names in Rajasthan: *deora, malvan, deorai, rakhat bani, oran, deo ghāt, mandir van*, and *bāgh*. The earliest writings in English about the sacred groves are by Dietrich Brandis, the first Inspector General of

Forests in India from 1864–83.[3] He wrote, "Though very few papers have been published on sacred groves, this does not mean that such areas do not abound in India." Commenting on the sacred groves of Rajasthan, particularly Rajputana and Mewar, he wrote that in Pratapgarh and Banswara such groves are common where people do not cut wood for personal use and only dead and fallen trees are removed for repair of temples or funerals.

Deep Pandey, a retired government forest officer, is representative of a whole wave of scholars interested in the religious, ecological, social, and economic aspects of sacred groves along with their conservation over the years through various traditions of indigenous resource management. Pandey studied sacred groves of the southern Aravallis for several years (1998). He collected their religious, ecological, social, and economic aspects and various traditions of indigenous resource management. The available resources, biodiversity, social beliefs, threats, and factors responsible for biodiversity depletion, economic status of village people, suggestions for conservation of sacred groves and joint forest management, were studied in the context of the sacred in nature. According to Pandey, more than 1,000 major orans are spread out in an area of more than 100,000 hectares. He notes that the sacred groves in Rajasthan are "the result of a complex ethno-scientific thinking of the local communities." Forests in hills reduce the runoff and help in groundwater recharge. The water thus becomes available in the *Bāwaḍī* (step-well) or pool located within the sacred grove during the lean months. Water also brings minerals and fertilizers in rich quantities. It is then logical that such resources are protected and conserved by the people. People might have institutionalized these arrangements during the course of time by attaching sacred value to it, to make collective management easy and long lasting.

In 2001, with the support from the Global Environment Facility of the United Nations Development program, KRAPAVIS (*Kṛṣi Avam Paristhitiki Vikās Sansthān*), an ecological and agricultural development organization, carried out a survey of the existing sacred groves in Alwar in the North-East of Rajasthan, as told to me by its founder Aman Singh. According to this survey, there were approximately 163 orans as of 2001. Villages with dense mountains had more orans than the ones with plains. They are also called *Devbanis* (literally, divine forests) in Alwar area. About 45 percent of them had an individual area ranging from one to three acres and about 14 percent of them had an individual area of about 30 acres with more than 100 species of trees (including *Neem* (Azadirachta indica), Pipal (*Ficus religiosa*), Khejadi (*Prosopis cineraria*), mango, lemon, banana, date, and guava). The study also found that each oran had at least one natural water source. About 31 percent had a pond, 17 percent had a waterfall, 38 percent had a well, and 14 percent had manmade boring wells. However, these traditional water resources were not well maintained. Orans also have 54 different species of animals (including deer, panther, tiger, hyena, boar, monkey, fox, wolf, snake,

[3] Dietrich Brandis, Indian Forestry, Oriental Institute, 1897, cited by Pandey (1998).

tortoise, mongoose, and chameleon) and 27 species of birds (including peacock, parrot, pigeon, crow, owl, ostrich, bat, swan, and crane).

From this study, I now describe some examples with noticeable success in community preservation efforts. The Shītal Dās and Gopāl Dās kī Deobani are two different orans located in two contiguous villages – Rainagir and Pehal, respectively – and are situated on a hill slope at a distance of three miles from each other. The total area under the orans is about 40 hectares. This area has been dedicated to deities Gopāl Dās and Shītal Dās that are worshipped by the community. These sacred groves are under the control of temples and managed by the village communities. Since the felling of the trees for timber has been prohibited, it has ensured that the fauna continues to survive up to the present time. Even today, this oran shelters rare plants and protects water sources in the area. Similarly, Bhaonta and Koylala are two nearby villages. After 10 years of successful forest protection effort by local communities of Bhaonta-Koylala, the forest was regenerated and it was declared a Bhairon Dev Lok Van Abhayāraṇya (people's sanctuary) in October 1998, named after Bhairon Dev, a local deity to which the forest is dedicated. This is a unique example of a local community's successful effort for conservation of biodiversity. With the regeneration of forest, villagers reported that a couple of leopards started frequenting the forest. The villagers claimed that the disappearance of tigers and other predators from the forest was the reason behind the depletion of forests. They mentioned that the presence of predators would inhibit people from going into the forest unless necessary and aid the conservation process (Shresth and Devidas 1999). In March 2000, India's President K.R. Narayanan presented the first "Joseph C. John – Down to Earth Award" for excellent community achievement in environmental management to Bhaonta-Koylala Grāma Sabhā (*Down to Earth*, April 2000).

Again, following Ann Gold, the above examples show a combined effect of faith of local communities working successfully with modern conservation efforts by the outside organizations. I now turn to my own fieldwork, which has several similar evidences of Bhil community's faith protecting their sacred grove. First, let me situate Bhils in the larger Hindu cultural history.

Bhils in the Hindu Culture and History

The religious ethos of Bhils sets the context in which they interact and protect some of their natural resources. Although the Bhils are called the "aborigines" of Rajasthan residing in the Aravallis, Roger Jeffery has problematized the "native" or "aborigines" labeling of them (1998). A noted folklorist Komal Kothari also noted that the problem is that the category of Bhils itself is much diffused (Bharucha and Kothari 2003).

Agreeing with the above observations, I would also like to suggest that the Bhils have transcended the dichotomy of "little" and "great" traditions. According to R.S. Mann, Bhils have been interacting with other Hindu castes for a long time (1993). The Bhagat movements propagated the "caste-Hindu" values such as vegetarianism, reincarnation, karma, notions of purity and pollution, fasting, and abstinence from alcohol. Among the major Bhagat movements to reach out to Bhils are the Māwajī (Vaiṣṇava), Govindgiri (also known as Lasodia), Rāma Deva, Nāthjī, Kabir, and Viśvakarmā. Through these various movements, popular Hindu deities that are now widely worshipped among Bhils are Mahādeva, Pārvatī, Rāma Deva, Viṣṇu, Brahmā, and Kṛṣṇa. Interestingly, as is the case with most Indian social groups, the Bhagat movements led to increasing social stratification in the Bhil community.

The Bhils adopted Bhagat symbols and activities such as marking of forearm, forehead, wearing rosary and saffron clothes, use of certain new speech symbols, sacred thread, and the mode of greetings. In the post-independence period, they also tend to be isolated both from the Bhils who are non-Bhagats as well as from the mainstream Hindu society. However, in the earlier Bhil–Rajput contact, instances of marriages between Bhil chiefs and the Rajputs were reported and other groups such as Bhilala, Garasia, and Patelia are assumed to have originated from Bhil–Rajput social relationships. The Bhils, like other Hindus, treat certain castes as socially superior, and others inferior. The caste Hindu influence among Bhils has also persuaded them to form their own reform groups such as Bhil Sudhārak Samiti that prohibits such Bhil practices as bride price and pre-marital sexual relationships. R.S. Mann notes two interesting groups among Bhils; Palia Bhils and Kalia Bhils. Some have treated Palia Bhils as original, pure or *Ujle;* and Kalia Bhils as impure or *Mele* (Mann 1993). As mentioned above, in the nineteenth century, Surmaldas and Govindgiri launched the Bhagat cult among the Bhils in an effort to "Hinduize" them. Hindu gods and goddesses were then widely adopted by the Bhils. As a major reformation process, both the Bhagat leaders inspired Bhils to stop consumption of meat and alcohol, to stop committing crimes and to lead "righteous" lives. The Bhagats distinguish themselves from Bhils by wearing white or saffron robes. Similarly, Bhils who convert to Christianity are segregated from mainstream Bhil community. Today, the Bhils are found in a variety of places, from remote forests to small towns and urban society, and can be generally considered a "Hinduized" caste.

The Bhils are adapted fully to the style of living which the territory demands, and are known for their bravery, carefree disposition, and excellence at archery. In stature, the Bhils are of medium to short height, generally slim, have a darkish brown skin complexion and wavy hair. They have been called *Vanaputras* and *Niṣādas* in the Sanskrit texts. The name Bhil may have been derived from Dravidian Bhil, Bīla, or Billa meaning bow. In Telugu and Kannada, the word for bow is villu that might have some etymological connection with Bhil also. Bhils were also mentioned in a 1240 CE Sanskrit work on dance *Bhillaveṣamupeyuṣim* used in Maharashtra and Karnataka. The community may have originated in the Dravidian-speaking southern part of India and then may have traveled north to its later home in the central Indian forests (Sumit Guha 2006: 108). They are the largest caste-group of Rajasthan and the third largest in India. They are concentrated in Udaipur, Dungarpur, Banswara, Chittorgarh, and Sirohi districts of Southern Rajasthan. They also live in Ratlam, Jhabua, and Nimar districts of Madhya Pradesh, in Panchamahal and Sabarkantha regions of Gujarat, and Khandesh region of Maharashtra. It is speculated that they were a roaming community in the central Indian region west of the Ganges before Aryans drove them into the forests. They also had small kingdoms in Gujarat, Malwa, and central India. In Rajasthan, they constitute approximately 39 percent of the total population. They were one of the first communities to have participated in the protest against the Indian Forest Act of 1878 that denied the village forest rights to indigenous people and sought to expand the commercial exploitation by the state machinery (Ramachandra Guha 2006: 96). Although the Bhil discontent did not manifest itself in Southern Rajasthan, it continued to be expressed through arson and other breaches of the forest law. Guha also cites Verrier Elwin, a pioneer of ecological anthropology who studied and lived with Bhils for several decades during 1940s and 1950s (Ramachandra Guha 2006: 100):

> Elwin showcases the intimate relationship between forests and the life-world of the adivasi. Elwin also argued that tribals had a deep knowledge of wild plants and animals; some could even read the great volume of nature like an "open book". They liked to think themselves as children of Dharti Mata, Mother Earth, fed and loved by her.

Prabhakar Joshi has mentioned about the sacred groves in Dungarpur district in Southern Rajasthan that are managed by the Bhil community (1995).

While D.K. Singh has noted the presence of the Rāmāyaṇa and the MBh on Hinduization by modern Hindu movements (2004), N.N. Vyas and others have noticed that both the Rāmāyaṇa and the MBh had already existed in Bhil and other communities of India since ancient times (Singh and Datta 1993). The main story of the Rāmāyaṇa remains the same with the Bhils. If we analyze the Bhil Rāmāyaṇa from the perspective of the anthropology of knowledge in terms of the interaction of the Hindu "great" tradition and the "little" traditions, the following interpretations can be made: rulers are seen as cruel, deceitful, and exploitative. Bhil life is sustained through the observance of rituals. Human subsistence lies in procreation that provides continuity to the society. Following Robert Redfield and M.N. Srinivas, Milton Singer (1972) explored the dichotomy of "great" and "little" traditions of India differentiating the "Brahmanical Hinduism" from "Local Hinduism." McKim Marriott (1965) pointed out that Sanskritization or Hinduization does not destroy or replace the local traditions, such as those practiced by Bhils. Sanskritization merely identifies the local deity with a universal one. One finds in an Indian village a transmutation and transformation of great and little traditions resulting from upward "universalization" and "parochialization." Frits Staal also reminds us that even the so-called "great" Sanskrit tradition itself originated from the "little" tradition and these origins are generally visible in a later stage. Agreeing with Staal, Sontheimer records examples of "little" traditions influencing the "great" traditions. According to him, tantric practices such as *śabarotsava*, in which the normal rules of hierarchy and sexual propriety are abandoned, and the *rāsa līlā*, in which the love of the gopīs for Kṛṣṇa is interpreted as emotional bhakti, both have their origins in the Bhil customs marked by sexual freedom (Sontheimer 1989). Vertovec also rejects the watertight dichotomy of great and little traditions (2000). Whitney Sanford also records that "little" practices in Braj are essential to the establishment and maintenance of "great" traditions and vice versa (2005). Frederick Smith in his study about the phenomena of possessions in Indian society concludes, "in folk contexts classical deities (and other figures) become folk, just as it may be shown that in classical contexts folk deities become classical ... [I]t fuels an argument for dissolving, or at least rethinking, the discussion between folk and classical" (2006: 151). He further mentions about "Vernacularization" and "Sanskritization" (2006: 595).

These observations match with the multiple versions of Hindu epics that are present in different parts of Bhil society as I showed above. Moreover, even the reverence for natural resources that I found in my fieldwork matched with the similar ethos found in other Hindu communities. Just as Hindus revere plants such as Tulasi (*Ocimum tenuiflorum*), Neem, and Pipal, Bhils have continued to protect their trees based on their faith in the deities surrounding them. I now present the examples of sacred groves that are maintained by Bhils in Southern Rajasthan. These examples lack the support from outside organizations and are based solely on native faith of the Bhils similar to the general Hindu tradition of revering nature based on local folklore.

Faith Based Bhil Sacred Groves

The groves in Southern Rajasthan are located near the village and close to a water source. Such groves are also at the top of small hillocks in the Aravalli mountain range, where people worship Bherujī, Bāvasī, and Mātājī. Khanpa Bherujī, Kukawas Bherujī, and Baḍī Roopan Mātā are examples of such sites in Udaipur. In my fieldwork, I visited several sacred groves near Banswara, Dungarpur, Chittorgarh, and Udaipur.

These sites have different forms of Hindu deities, such as Hanumān, Durgā, and Śiva. The presence of the temples in these groves has protected their trees. People do not harm sacred groves mainly because of their traditions, believing that those who cut or use an axe in a sacred grove may be harmed by the presiding deity. I now describe some such sites.

A legend circulates in the Madar village in Udaipur about the Ekpaniyā Bāvasī sacred grove. Several decades ago, somebody wanted to cut a Haldu tree from the forest. From the first cut, milk flowed down, and water in the second cut. The third cut yielded blood and the axe-man lost his sight. His sight could only be regained when the axe-man promised to construct a new temple for

Ekpaniyā Bāvasī. These beliefs might have strongly influenced conservation of sacred groves. Continuous community protection of sacred groves has resulted in several large trees. For example, there is a large tree of Churail (*Holoptelia integrifolia*) growing in Amrakjī sacred grove. This is the largest tree of this species in India, having a height of more than 33 meters, and a girth of 6.91 meters. Only fallen and ripe fruits are collected from the grove and the wood from mature trees is used only in religious rituals. The water sources found in this and other similar groves provide for drinking and irrigation (Joshi 1995).

Another sacred grove near Udaipur is called Ubeśvara Mahādeva with a temple dedicated to Śiva. Since it is situated near a water stream, it also serves as a watering and resting place. The cowdung collected from the nearby cattle is allowed to dry and the dried dung cakes are used to cook *bāṭī* (spherical traditional bread) by villagers and pilgrims. The arrangement ensures the sanctity of the grove and provides ample stock of fuel. Sacred groves also provide meeting places for the community to discuss socio-economic issues and to resolve their personal grievances. Temple forests are managed and maintained to serve the temple. This may include religious, economic, and social functions. They are managed to meet the requirement of temples that in turn support the social functions. Similarly, the Śrīnāthjī temple near Udaipur has a sacred grove located in Gautameśvara forest block. The temple management does not derive authority from the state forest regulations and protects its grove against grazing, fire, and illicit felling (Joshi 1995).

Paoti Koda is a site at the Ved Forest Block near Mewara. The most common trees here are Bibhitaki (*Terminalia bellirica*), Arjuna (*Terminalia arjuna*), and Gum Karaya (*Sterculia*). Among these arboreal species, several herbs, shrubs, and vines form a dense mat. In this grove, there is a natural water source. Water trickles constantly from under the miniature cave beside the trees where a saint named Daljī Bhagat died several years ago. On the anniversary of the saint's death, people from nearby villages gather there to perform the rituals. Throughout the night, they sing and dance, in typical community frenzy, and depart in the early hours of the dawn silently. The water falls at a height of about 50 feet. It is said that at the source some six feet below the top of the cliff was a huge rock projecting out as a platform on which water flowed to drop at the center of the shallow pool 50 feet below. Several years ago, while a mother was hewing firewood in the surrounding forest, her daughter Varju fell on this rock from where she slipped in water on the angled rocks. Another child, too afraid to move, stood at the spot dazed and shocked. On discovering the absence of Varju, a frantic search was carried out in the woods around but to no avail. Accidentally, some time later she was discovered by the Bhils sitting in the grove below. From that day, the place gained a reputation as a pilgrimage center attributed to the beneficent soul of the departed saint – Daljī Bhagat. Over the course of time the rock was thrown down by the fury of a storm and the portion that was held by the cliff has opened into a cave. The remaining grotesque big stones in the cave are still worshipped as guarded by the spirit of the saint who had saved the girl.

Figure 5.1 View from inside a Bhil temple

Figure 5.2 A Hanumān temple inside a Bhil sacred grove

Figure 5.3 A Durgā temple inside a Bhil sacred grove

The place, Paoti Koda, with the cave, water stream and the untouched old and majestic trees is conserved in the ample woods of the Ved Forest (Joshi 1995). There is an abundance of such stories connected with different groves.

Boreśvara is a site at the Phalior forest block in Aspur tehsil. On the banks of the river Mahi, skirted at the river end with rows of Negundo (*Vitex negundo*) shrubs is the Boreśvara Mahādeva sacred grove dedicated to Mahādeva, a form of Śiva characterized by the abundance of arboreal fauna and flora. The most common species of the grove are Arjuna, Tendu (*Diospyros melanoxylon*), Gum, Bel (*Aegle marmelos*), Cluster Fig (*Ficus racemosa*), True Date Palm (*Phoenix dactylifera*), Kadamba (*Anthocephalus cadamba*), and Fig (*Ficus carica*). The trees in the grove are older and taller than those in the outside forest, with several black langurs, squirrels, and birds in the dense canopies. Its situation from a distance can easily be perceived by a distinct arboreal patch rising over the surrounding teak forest with average and smaller sized trees. The route to Boreśvara is rich in Cassia draped with the needle-leaved Shatavari (*Asparagus racemosus*), and *Javanicus* climbers.

Village Mālvan is a site near Rangela. The most common trees here are Tendu (*Diospyros melanoxylon*), Bibhitaki, and teak (*Tectona grandis*). The shrine is a permanent construction with hung mango leaves constituting the toran and poles of teak bearing the flags in front. Another village of the same name is near

Charwara. This grove is comprised of about 20 specimens of Pīpal and Jasmine (*Jasminum abyssinicum*). As in the previous case, flags hoisted on teak poles were arranged at the front of the shrine. Basil (*Ocimum basilicum*) was cultivated in the compound of the grove. At several places in the Jhadol Forest of Udaipur district, groves exclusively of bamboo are seen. The bamboo clumps growing in these groves are old and tall (Joshi 1995). Unlike the above-mentioned groves where the local beliefs have been the main driving force in their preservation, in my following examples governmental efforts have also joined to take advantage of the local beliefs in preserving the natural resources.

Sacred Groves Based on Faith and Governmental Efforts

There are two trees, known as the Indian Kalpavṛkṣa (*adansonia digitata - baobab*), that are present at about three miles from Banswara city in the Anand Sāgar near Mahi River. Bhils call the trees *Kalpadeva* (lit. "tree-deity") and they are declared as national monuments by the Indian government. Of the pair, the larger one, 20-feet high, is designated as *Rānī* (queen) and the smaller, 14-feet-high, tree is designated as *Rājā* (king). Apparently, another tree nearby was submerged in the Mahi River a few years ago. The Bhils have immense faith in these trees and travelers never miss the opportunity of having a *darśana* of the trees. People desirous of having any wishes fulfilled circumambulate both the trees, keeping their wishes in mind, contemplating a return in order to perform *yajña*, the fire sacrificial ritual, as a token of their gratitude. By the number of such rituals being performed, it appears that the wishes of several Bhils have been fulfilled. Even during birth and death, the Bhils worship the *Kalpadeva* with specific rituals. Newly married couples tie the sacred threads around both the trees 108 times, believing that like the thread they will be tied in matrimony for eternity.[4] A newly born baby is carried around *Kalpadeva* anticipating a long life. On *Hariyālī Amāvasyā*, a festival night in rainy season, the barren place becomes alive with people and a fair is held - attended by Bhils from remote places. The trees are given a "bath" with the Mahi water and people return with soaring hopes. The dried bark and fruit of the tree are placed within granaries by farmers in hope of bringing prosperity. The age of these trees may be over 500 years according to a local forest officer. Similarly, Mahua (*Madhuca longfolia*) is also considered sacred by the Bhils.

To restore the sacred groves of the Aravallis, a program called Aravalli Devvan Sanrakshan Abhiyan (Aravalli Sacred Grove Conservation Campaign) was launched in 1992. This included protection of groves, planting of indigenous

---

4 In Uttar Pradesh's Etawah and Mainpuri regions, a newly married couple is shown a pair of cranes. Sighting a pair of the cranes bodes well for marital bliss, they believe. The sarus's conjugal devotion is celebrated in the folklore and the classical literature (*Down to Earth* Magazine, April 15, 2007).

species, soil and water conservation, and a participatory approach to restoration. Some of the restored groves include Moria kā Khuna, Jhameśvarajī, Amarakjī, Ubeśvarajī, Dhinkli, Haldu Ghati, Banki, Khokhariya kī Nal, and Ambua sacred groves (Pandey 1996).

Another major government managed environmental project, which is based on the religiosity of Bhils, is Sītāmātā Wildlife Sanctuary, located 30 miles from Chittorgarh and 60 miles from Udaipur in the hilly area of the Aravallis. Its area is about 400 square kilometers. It was declared as a sanctuary in 1979. In this sanctuary, around half of the trees are teak. Besides these, salar, tendu, amla (*Phyllanthus emblica*), and bamboo are also present. The leopard, hyena, jackal, fox, jungle cat, porcupine, spotted deer, wild bear, four-horned antelope, and Nilgai are the animals found here. The most conspicuous animal of the Sītāmātā sanctuary is the flying squirrel, usually gliding from one tree to another after sunset. Its activities are nocturnal and, during the day, it hides in the hollow. According to a 2002–4 study, 243 genres belonging to 76 families are used by the Bhils of about 50 villages around the Sītāmātā sanctuary as the means of primary health care to cure various ailments. The study revealed the new ethnobotanical uses of 24 plant species belonging to 20 genres (Jain et al. 2005: 143–57). The major myth associated with this site is that Sītā lived here at the hermitage of Vālmīki, after being exiled by Rāma in the *Rāmāyaṇa*. She also gave birth to her two sons, Lava and Kuśa, here.[5]

There are trees at this site that are associated with the adventures of both her sons, such as their defeating the army of Rāma. Two waterfalls are also named after them. The highest point that pilgrims visit in this site is a temple dedicated to Sītā that is located in a huge valley formed between two mountains. It took some adventurous mountaineering for me to reach this highest point. Local Bhils believe that Sītā ended her life in this valley by taking shelter in the earth. There is also a temple dedicated to Hanumān where supposedly Lava and Kuśa had tied him during their encounter with Rāma's army.

I encountered similar mythic associations at several places in Southern Rajasthan. The most popular such temple is *Ghoṭiyā Ambā*.

According to folklorist Mahendra Bhanavat (1993), it is widely believed that the Pāṇḍavas in their years in exile visited here, and from the seeds of their mango fruits that were gifted by Indra, grew mango trees in the temple area. According to a folktale, 88,000 sages were invited here for a special treat with mango-pulp in the presence of Kṛṣṇa. Banana leaves were used as plates for that special party and those trees in the area currently do not grow fruits. It is believed that the rice plants grew from the grains that fell off from the food served in that party. On the *Kṛṣṇa Amāvasyā* of Hindu month of *Caitra*, there is a large fair held here in which Bhils from Rajasthan, Gujarat, and Malwa assemble. Pilgrims collect the upper portions of the rice plants and store them in their granaries back home. It is believed that such houses never face drought and enjoy

---

5 This tale appears in the seventh book of the Rāmāyaṇa titled Uttarakāṇḍa.

Figure 5.4 A sign at the Sītāmātā Sanctuary showing the birthplace of Sītā's two sons

steady growth of grains. In the surrounding area of about 40 square kilometers is a dense forest of mainly sagwan trees. According to the local folklore, Rāma visited the Rāma Kuṇḍa during his exile, the Pāṇḍavas visited the Bhīma Kuṇḍa during their exile, and they had used a tunnel as their passage during the rainy season that connects the Bhīma Kuṇḍa to Ghotiyā Ambā.

I now describe the relatively new ritual of sprinkling saffron, *Kesar Chhāṅṭa*, which has helped protect several areas of forests in Southern Rajasthan. People collect saffron from a nearby temple and sprinkle it collectively around natural forest patch; this puts voluntary restriction on green felling. Forests are thus treated as de facto sanctuaries being maintained and protected by people living in and around the forest. This not only protects the forest against the green felling, but also allows birds and wild animals to roam freely in the area. Management practices in *Kesar Chhāṅṭa* forest include protection, patrolling by a community appointed security guard, restrictions to green felling, equitable gathering of dead and fallen wood, harvest of non-timber forest produce and grasses. These areas represent the ultimate example of social fencing, i.e., a boundary created by a neighborhood to bar people from entering the nearby forest areas. People punish themselves for any act not permitted by *Kesar Chhāṅṭa* tradition and thus sprinkled saffron serves as a social barrier. This also protects water

Figure 5.5 Temple at Ghoṭiyā Ambā

sources, medicinal herbs, and fruit-bearing species. For example, in Udaipur South Forest Division about 12,000 hectares of forests are being protected by *Kesar Chhāṅṭa.* These areas include Vijaya Talai in Salumber, Alsigarh, Bada Bhilwara, and Shyampura in Udaipur (West), Madri in Jhadol, and Nayanbara in Khairwara. This faith has also been utilized by the Forest Department to initiate the protection by local communities at several places. Sagwara, Kojawada, Vijaya Talai, Pargiya Pada, Ranjitpura, Jagnathpura, and Barli Padi forests are protected and managed by the people. People in many of these areas were motivated and organized by Seva Mandir, a leading NGO in India working in Udaipur district. The usual practice followed is that people collect Kesar from a famous temple Rishabhdev-Kesariyajī and move around the forest area beating the drums to communicate the message that Kesar is being sprinkled and the area has come under community protection (Pandey 1996). As of 1995, through its work on Natural Resource Development, Seva Mandir has treated 12,343 hectares of degraded common and private pastureland. It has also worked on the protection, plantation, and management of more than 1,000 hectares of forestland under the Joint Forest Management scheme, in collaboration with the Forest Department. It has completed the watershed treatment work on 8,959 hectares of land spread across 27 sites. Seva Mandir has also constructed three masonry anicuts (water harvesting structures) and helped to establish 33 community managed lift irrigation systems. Through its agricultural extension work, Seva Mandir has helped farmers establish more than 1,000 vermi-compost units and hundreds of horticulture nurseries. The organization has also conducted close to 100 animal camps.[6]

Another sanctuary that I visited was Tal Chhapar Sanctuary in the Sujangarh Tehsil of Churu district. It lies on Nokha–Sujangarh state highway and is about 50 miles from Churu and 80 miles from Bikaner. It provides a unique refuge for blackbucks. Tal Chhapar Sanctuary, with almost flat tract and interspersed shallow low-lying areas, has open grassland with scattered *acacia* and *prosopis* trees that give it an appearance of a typical savanna. The word "tāl" means plain land. The rainwater flows through shallow low-lying areas and collects in the small seasonal water ponds. The geology of the zone is obscured by the heavy winds. Some small hillocks and exposed rocks of slate and quartzite are present in the western side of the sanctuary. The area between the hillocks and the sanctuary constitutes the watershed area of the sanctuary. The whole sanctuary used to be flooded by water during the heavy rains but with salt mining going on in the watershed, hardly any rainwater falling on the hillocks reaches the sanctuary. During the British rule, Mahārājā of Bikaner managed Tal Chhapar as a *Shikāragāha*, hunting ground, and ample protection and facilities were provided to attract and maintain a sizeable population of wild animals in the area. Tal Chhapar with its rich diversity of animals and birds was one of the favorite places of game reserve for the rulers of Bikaner state before independence. I found similar reserves for the rulers of Dungarpur and Banswara in Southern Rajasthan, just as Ann Gold found in Ajmer district in central Rajasthan (Gold and Gujar 2002).

In general, I met people, including government forest officers, who recalled the pre-independence days of kings with dense forests, just as Ann Gold found in Ajmer (Gold and Gujar 2002). This however is different from Ramachandra Guha's observation about Tamilnadu (2006: 96). Guha finds from his archival research that almost all the land was available to the public at large in Madras state in the nineteenth century and there was no royal control on the forests there. However, he cites Dietrich Brandis, the first Inspector General of Forests in India from 1864–83, who appreciated the forest conservation by Rajasthani Rajput kings, "as a good example that the forest officers of the British government would do well to emulate" (2006: 106).

Concluding Remarks

I presented two kinds of Bhil sacred grove above. While the former ones were solely based on the indigenous Bhil beliefs and folklores, the latter also included the active participation of outside agencies and the government. In both the

6 www.sevamandir.org/Strengthening%20livelihoods.htm (viewed on June 15, 2007).

categories, the Bhil religious values are similar to the larger Hindu values of regarding the natural resources based on ancient religious myths and tales. Thus, I suggest that Bhils can be situated in the larger Hindu culture of India, which has several examples of faith in the natural resources such as the Himalayas, the Ganges, and several gods and goddesses with animal features. The reverence for natural resources that I found in my fieldwork matched a similar ethos found in other Hindu communities. Just as Hindus revere plants such as Tulasi, Neem, and Pīpal, Bhils have continued to protect their trees based on their faith in the deities surrounding them. However, like Swadhyaya movement, their environmentalism is limited to the confinement of their groves. They have not been able to conserve the forests outside their sacred groves. Thus, the evolution of religious ethos into ecological ethos that I found in Bishnois seems to be missing in the Bhils and in Swadhyaya movement (as it exists today). Unlike Bishnois, Bhils and Swadhyayis are not seen protecting the ecological resources outside their sacred sites. Whatever limited environmentalism of Bhils that I found around their sacred groves is still different from the environmentalism that we find in Western organizations and Indian modern organizations that I intend to describe in the next chapter.

Chapter 6

Modern Organizations Adapting to Ecology

For Indians, preservation [of their traditions] is also an act of responding to the West. In modern times, responding to the Western presence and the global phenomenon of Westernization is no longer a matter of personal choice or predilection. Even withdrawal and silence, and affirmation and continuation of traditional forms, are ways of responding.

Halbfass (1988)

As a comparison to my Indian case studies, I now present my observations on some American and other modern Hindu and Jain organizations working on ecological issues. To situate my ecological research about Rajasthan in a comparative context, I was looking for similar examples in New Jersey when I discovered GreenFaith, the "interfaith partners in action for the earth."[1] I met the executive director of GreenFaith, Rev. Fletcher Harper, at his office in downtown New Brunswick. Although they have primarily worked with Judeo-Christian organizations so far, their mission is "to inspire, educate, and mobilize people of diverse spiritual backgrounds to deepen their relationship with the sacred in nature and restore the environment for future generations." Harper told me that he is interested in expanding his network to include Asian community organizations in New Jersey. He was ordained as an Episcopal priest in 1992 after attending Union Theological Seminary and serving as a chaplain in a cancer ward. He served for 10 years as a rector of two parishes in northern New Jersey and was active in denominational leadership. I was interested to note his Christian background with his current role in the field of environmental activism. He provided me with several press briefs and photographs in which he and others from GreenFaith approached different religious organizations to educate them about environmental issues. Several Christian sects present some sort of theological limitations in terms of viewing the environment as potentially divine. In her 2007 article, A. Whitney Sanford cited Klassen to note that the Western religious discourse stemming from Judaism and Christianity has emphasized experiencing the divine as transcendent rather than as immanent within the material realm.[2] However, tribes of Rajasthan that

[1] www.pluralism.org/research/profiles/display.php?profile=73731 (viewed on January 15, 2007).

[2] Similarly, Piers Vitebsky noted in his ethnography of the Sora tribe in Andhra Pradesh how its ecology was seriously affected when the majority of this tribe converted from their ancient indigenous traditions to Christianity (1998). Such interactions of different traditions and communities have also been occurring in India for several millennia, especially among the so-called "little traditions" and "great traditions."

I presented above see no theological dilemma in regarding natural resources as divine, even while the "Hinduization" process has continued there led by different Bhagat groups. Anne Feldhaus suggests that the variety of idioms and other religious phenomena in India is much greater than any strict dichotomy of "great" and "little" could comprehend. These different idioms often can be used to say surprisingly similar things about the seemingly diverse Indic traditions (1995: 16). Thus, although the Western religions present a theological dilemma in regarding nature as divine, they are transcending this dilemma and are coming to the forefront in raising and spreading ecological awareness. This "renewal" of Western religions far outweighs their Asian counterparts. Let me show some examples below.

The Roman Catholic Church has issued calls to save and protect the environment in its various appeals issued on different occasions by the Pope (*The New York Times*, December 25, 2007). According to a *Washington Post* report, "The Greening of Evangelicals,"[3] several Christian leaders are now working for environmental issues. In this report, John C. Green, director of the Ray C. Bliss Institute of Applied Politics at the University of Akron, noted that polling has found a strengthening consensus among evangelicals for strict environmental rules, even if they cost jobs and higher prices. In 2000, about 45 percent of evangelicals supported strict environmental regulations, according to Green's polling. That jumped to 52 percent in 2005. I already presented the efforts of GreenFaith which is at the forefront of spreading ecological awareness in New Jersey.

Coming back to Indic traditions, a majority of rural Hindu communities and castes are still living their *premodern* lives and therefore they are yet to wake up to *postmodern* environmentalism (although my case studies show some exceptions). However, the modern urban Hindu organizations have included environmentalism in their agenda just as their Christian counterparts have done. When I started looking for similar ecological initiatives in other neo-Hindu movements and organizations, I was surprised to find that several movements and organizations are striving to "become green." As Halbfass notes in his quotation above, most of these new organizations are responding to modern ecological problems with modern means much in line with the

Tomalin has noted similar deterioration of sacred groves when tribal communities had to adopt mainstream Hinduism.

3 www.pbs.org/wnet/religionandethics/week920/cover.html (viewed on January 30, 2007). Christian leaders that have responded to ecological issues include Rev. Leroy Hedman (leader of Georgetown Gospel Chapel), Rev. Ted Haggard (president of the 30-million-member National Association of Evangelicals), James Dobson (Leader of Focus on the Family), Chuck Colson (Leader of Prison Fellowship Ministries), Rev. Richard Cizik (vice president for governmental affairs, National Association of Evangelicals), *Christianity Today* magazine, Larry Schweiger (president of the National Wildlife Federation), and Rev. Jim Ball (executive director of the Evangelical Environmental Network).

Western style of activism. Below I present some of the stated ecological missions by several organizations.

One of the most widely known neo-Hindu movements, Ramakrishna Mission, founded by Swami Vivekananda, is already known for its Protestant style social work. In addition, Kamala Chowdhry describes its 37 Forest Protection Committees in West Bengal (Paranjape 2005: 139). These have helped stop the pilfering and illicit felling of trees. The members of these committees, mostly wage laborers and pastoralists, share the benefits of forest protection themselves. Another global Hindu organization, BAPS Swaminarayan Sanstha, declares several environmental initiatives it has undertaken, such as seven million aluminum cans and 5,000 tons of paper collected for recycling, 1.5 million trees planted in 2,170 villages, 5,475 wells recharged in 338 villages, solar energy and biogas used at its temples, and 497 rain harvesting projects completed. For their various livestock projects, BAPS cattle farms have been awarded 34 National Livestock Awards.[4]

One of the newest global Hindu movements, Art of Living, headed by its founder Sri Sri Ravi Shankar, has developed biodynamic farming propagated by Sri Sri Mobile Agricultural Institute. It has trained the farmers to revert to organic farming, to plant more trees, and to adopt soil and water conservation measures. It particularly promotes the introduction of organic farming techniques, such as vermi-composting and the use of natural pesticides, biofertilizers, and effective microorganisms.[5] Similarly, the All India Movement for Seva, founded and currently headed by Swami Dayananda Saraswati, has included water management and planting campaigns in the Indian states of Tamil Nadu, Karnataka, Andhra Pradesh, Orissa, and Maharashtra.[6] Another recent Hindu guru, Amma, "The Hugging Saint," has established GreenFriends. It engages in tree planting and the maintenance of plants. Its members also practice eco-meditation, a method of re-establishing the vitally important harmony between nature and humanity. Through the *Amrita Vanam* (Amrita Forests) Project, GreenFriends members undertake large-scale forestation projects in conjunction with state forestry departments. Every November, GreenFriends distributes and plants 100,000 saplings in the state of Kerala. GreenFriends also aims to restore the lost tradition of the Kerala manor garden, comprising a grove, a pond, and a shrine.[7] Ecological initiatives by the Brahma Kumaris World Spiritual University (BKWSU) are focused on renewable energy at their headquarters in Mount Abu, Rajasthan. Their research and development program comprises the following technologies: hybrid alternative energy systems, passive solar architecture,

[4] www.swaminarayan.com/activities/environmental/index.htm (viewed on March 30, 2007).

[5] http://artofliving.org/Service/Community/BioDynamicFarming/tabid/118/Default.aspx (viewed on March 30, 2007).

[6] www.aimforseva.org/projects/environment.html (viewed on March 30, 2007).

[7] www.amma.org/humanitarian-activities/nature (viewed on March 30, 2007).

Photovoltaic power packs, solar hot water plants, solar steam cooking systems, and water recycling technologies. They have also collaborated with various Indian and European agencies sources to conduct their research.[8] In 1995, American Sai Organization, an establishment by Sathya Sai Baba, launched a program called "The Earth – Help Ever Hurt Never." The list of the projects involved reusing or recycling the batteries, eyeglasses, junk mail, papers, shopping bags, greeting cards, and shoes. In addition, it promoted the vegetarian diet and launched a tree planting campaign.[9] Another neo-Hindu movement, Gayatri Parivar, organized a river-cleaning operation on the banks of Har kī Pauri, Haridwar on October 26, 2005. The Governor of Uttaranchal, head of All World Gayatri Parivar Pranav Pandya, and students, teachers, and volunteers of their various institutes removed 50 trucks of waste material from the riverbed.[10] On another occasion, in Khammam district, Andhra Pradesh, Gayatri Parivar invited Sunderlal Bahuguna to spread the message in 24 local schools to celebrate the Holi festival without burning the trees and bushes. There is an increasing focus to celebrate other Hindu festivals in an eco-friendly way.

In Orissa, the Sacred Gift builds on the people's devotion to Lord Jagannath – a devotion that has been a key element of Orissan culture for at least 2,000 years – and aims to set up three forest conservation zones, each incorporating about 10 villages sited in state-owned forestlands. Since 2000, each village has had a Forest Protection Committee to promote joint forest management based around practical incentives and employment schemes. In 2001, the local communities developed a management plan in collaboration with Alliance for Religions and Conversation. By mid-2007, 2,369 hectares were earmarked for plantation under the Shri Jagannath Vana Prakalpa Forest Project.[11]

In my discovery trip of environmentalism in Rajasthan, I was struck by a unique center for wildlife protection inspired by the Jain tradition, Shree Sumati Jeev Raksha Kendra,[12] located adjacent to the town of Malgaon in Sirohi district. This campus was developed by K.P. Sanghvi Group and it comprises a Jain Temple Complex and Animal Welfare Center. The center, established in 1998, takes care of sick, injured, old, retired, homeless, and rescued stray cattle including cows, buffalos, dogs, and donkeys.

The Institute has a *Gośālā* (cow-center) that is spread over more than seven million square feet that takes care of more than 5,000 stray cattle. The center employs more than 150 persons to look after the cattle and three veterinary doctors to give medical aid to the cattle. Cow milk is used for rituals at the

---

8 www.bkwsu.org/whatwedo/globalinitiatives/environment.htm (viewed on March 30, 2007).

9 www.sathyasai.org/organize/z1reg01/tehehn/tehehn.html (viewed on March 30, 2007).

10 www.dsvv.org/?Social-Services/Ganga-Safai (viewed on May 15, 2007).

11 www.arcworld.org/projects.asp?projectID=337.

12 www.pavapuri.com/kendra_intro.htm (viewed on June 15, 2007).

adjacent temple complex and the garden in the shelter premises provides flowers for the temple. Mr. Ramavtar Aggarwal, secretary of the All India Gośālā Federation, said that there are more than 3,000 Gośālās in India and Sumati Center at Pavapuri is one of the biggest in India. Another organization called Love4Cow maintains a nationwide list of Gośālās maintained by Hindus and Jains and lists more than 670 Gośālās in Rajasthan alone. Though not trained as a scholar of Jain tradition, Michael Tobias recognized a commonality between his own environmental interests and the Jain worldview (1991). Similar efforts by the Hindu and Jain communities are widely reported from many places in India (Lodrick 1981). *The Hindu* reported a Jain/Hindu Gośālā Satyam Śivam Sundaram Gaunivas at Gaganpahad near Hyderabad (July 5, 2005). Considered South India's biggest cow shelter and managed by the Shiv Mandir Goshala set up by jeweler-turned-philanthropist, Dharam Raj Ranka, the shelter houses over 2,000 cows rescued from slaughterhouses in addition to 200 bulls. Justice Gumanmal Lodha, a Jain ex-lawmaker from Rajasthan, during his tenure as the chairperson of the National Commission on Cattle, published a detailed report to ban cow slaughter in India and submitted it to the Union Government of India. The report, in four volumes, called for stringent laws to protect the cow and its progeny in the interest of India's rural economy. Lodha moved close to a national ban on cow slaughter in India, although most states except Kerala have already banned it long ago. However, this political activism is also interpreted as pseudo-environmentalism since it is tied with a bigger motive of luring the "Hindu" vote-banks in the electoral politics of Indian democracy.

Another dimension of Jain principles in practice is evident at the Jain Bird Hospital in Delhi at the Digambar Jain Temple, opposite Red Fort near Chandni Chowk, where the only patients admitted are birds, preferably the vegetarian ones (Lodrick 1981: 17). It was established by Prachin Shri Aggarwal Digambar Jain Panchayat in 1956 on the Jain principle of aversion to killing. The hospital has separate wards in the form of cages for different species like sparrows, parrots, domestic fowls, and pigeons. It also has a research laboratory and even an intensive care unit for its serious patients. The nearby people, especially the Jain merchants, bring the birds for treatment that are usually wounded by fowlers, ceiling fans or by other means. The hospital admits a maximum of 60 injured birds per day and about 15,000 in a year. They are treated, bathed, and fed a nutritious diet for their fast recovery. Later, the birds are set free from the hospital's terrace overlooking the Red Fort. To show yet another example of Jains protecting the animals, in 1969, Goa's largest wildlife sanctuary was named after Mahāvīra. The governor at the time was a Jain and he suggested the name. In 1982, the local Jains donated 12.5 million rupees for the development of the Sanctuary.[13]

Continuing my description of the environmentalism of modern religious organizations, in North America, JAINA (an umbrella organization of several Jain

[13] I thank Shonil Bhagwat, Oxford Long-Term Ecology Laboratory, for this information.

associations) has a Jivadaya Committee that has selected about a dozen animal shelter organizations in India. JAINA encourages its Jain members to send donations to these organizations to help fund animal shelter and protection activities in India. In 2003, Nitin Talsania, active member of several JAINA committees in New Jersey, helped the Voith family of Angelica, New York, in their struggle to raise cows in accordance with their Hindu practice.[14] Similarly, the Jain Center of Southern California joined hands with "The Purple Cow and Friends," a non-profit organization near San Diego. Its director Ms. Tiffany and her co-workers collected fruits, vegetables, and grain from local farmers and grocery stores for more than 100 animals (including cows, dogs, pigs, goats, horses, and other birds).[15]

In July 2007 at the 14th biennial JAINA convention in New Jersey, I met Saurabh Dalal, president of the Vegetarian Society of the District of Columbia, the oldest vegetarian society in the USA. Inspired by his Jain background of *Ahiṃsā*, non-violence, Saurabh has been active in spreading the awareness about the connection between meat based diet and global warming. I attended his talk at the convention in which he showed several interesting facts. A vegetarian diet requires only 10 percent of the land that the standard meat-based diet does while a vegan diet would require only 5 percent of the same land. Similarly, by becoming vegan, one can save 3,900 gallons of water.[16] I also met a Jain activist, Pravin K. Shah, who became a strict vegan after his visit to a dairy where he saw the cruel treatment to the cattle there.

Another relatively newer oasis, "Tapovan Ashram", is in the denuded forests and hill region of the Aravallis at Naya Kheda village in Udaipur district. This is a lush green orchard of about 15 acres of land with a high water table and exotic herbs and fruits. This began as an experiment by a Jain horticulturist Ratan Chand Mehta in 1991. He harvested the rainwater from the nearby hills by *tatbandi* and soon the water table of that small area was higher than the surrounding areas. The farming here is titled "*Bhagwan Bharose Mast Kheti*" (God Blessed Bio Diverse Farming). I asked Mehta about his inspiration for the Ashram. He emphasized his Jain background and mentioned the Ācārāṅga Sūtra, one of the earliest texts that recorded Mahävir's teachings. Christopher Chapple also cites Ācārāṅga Sūtra (2.4.2.11–12) in which Mahāvīra tells his disciples to recognize the inherent value of trees and to turn their thoughts from materiality by reflecting on the greater beauty of sparing a tree from human exploitation (2002: 136). Sadhvi

---

[14] www.nomoreinjustice.org/media_articles_cowscantcomehome.htm (viewed on May 3, 2007).

[15] www.jainworld.com/society/jainevents/GJE2003/Purple%20Cow%20Animal%20 shelter.htm (viewed on May 5, 2007).

[16] Such arguments seem to be vindicated by the latest study by the National Institute of Livestock and Grassland Science in Tsukuba, Japan (*Telegraph*, UK, July 23, 2007). Similar conclusions were reached by Gidon Eshel and Pamela Martin, assistants of geophysics at the University of Chicago (ABC News, April 19, 2006).

Shilapi also cites from Ācārāṅga Sūtra (1.1.5) to show that Mahāvīra proclaims that anyone who neglects or disregards the existence of earth, air, water, and vegetation disregards his own existence that is intrinsically bound up with them (Chapple 2002: 160).

Jainism classifies the various living beings under different grades according to their development and sense faculties. Living beings fall under two broad classes, *trasa* or mobile and *sthāvara* or immobile. Trasa beings are those that possess two, three, four, and five sense organs. Sthāvara beings are those that have only one sense organ of touch, and they are of five kinds, earth-bodied, water-bodied, fire-bodied, air-bodied, and vegetables. A Jain monk is supposed to avoid injury to all trasa and sthāvara beings. A Jain householder is also supposed to avoid injury to trasa beings and is supposed to minimize injury to sthāvara beings. Therefore, Jain householders avoid eating meat and those vegetables that are roots or trunks of the plants. They also avoid eating fruits and vegetables that may contain living organisms such as figs and honey.

Evidently, Jain history is full of examples, legends, and tales of protecting and avoiding injury to plants, animals, and environment in general (Chapple 1998a). Jains believe that the nineteenth Jain Tīrthaṅkara Mallinātha had taken the responsibility of protecting the forests. Anne Vallely argues that Jains in the North American diaspora have made a shift from the conservative and orthodox Jain community in India (2002a). She notes that most Jains in India are not so active in social and ecological areas, whereas the "diaspora" Jains, especially the second-generation youth, are active in interfaith and animal welfare forums and groups. However, my observations from the fieldwork in India do not match her conclusions. In my case studies of animal welfare and protection, the Jain community has taken active role together with other communities of India. This also matches with the observation of Peter Flügel who notes the "sociocentric" role of Jain community in India (2005).

Concluding Remarks

From all the above-mentioned examples, we find that several modern Indian and Christian organizations have begun adapting to the environmental problems in their own ways, reflecting Orr's prediction. The lived reality of Christian leaders today seems to have successfully evolved over last few decades in the United States to include environmentalism in their agenda. On the other hand, although Indic traditions already present reverential examples of several trees, mountains, rivers, and other natural resources in their scriptures and rituals, only recently new movements have been able to connect the rituals with environmentalism as evident from the above examples.

Although the modern organizations have already begun "becoming green," their ecological initiatives are largely driven by their awareness about ecological issues. Their environmentalism is a conscious *response* to the problem. On the other hand, Bhils, Bishnois, and Swadhyayis have continued to practice their traditional lives not as a *response* to the modern ecological problems but because of their inspiration driven from their myths, legends, and teachings by their gurus. Most of the modern organizations that have included environmentalism as a distinct category in their agenda are also modeled as a "New Religious Movement" rooted on "religion" as a distinct category. Since the environmentalism of Swadhyayis, Bishnois, and Bhils appears different from the modern organizations, I would like to theorize my case studies using the Indic concept of dharma. I would like to call the environmental practices of Swadhyayis, Bishnois, and Bhils as Dharmic Ecology. Rather than motivated by the modern scientific awareness of ecological issues, Dharmic Ecology reflects the traditional ecological awareness of these communities which have been more effective than governmental initiatives, as I noted before. To better interpret and explain Dharmic Ecology, I now present a survey of dharma and its implications for ethics in the next chapter.

Frederique Apffel-Marglin and Pramod Parajuli have argued that despite the recognition of bio-divinity in the Hindu texts and traditions, "Hinduism (or any other religion) cannot offer the solution for contemporary environmental crisis" (2000). My survey of modern organizations above seems to concur with this assertion. Most Christian and Hindu religious organizations had to adapt themselves to the current environmental issues. Without their conscious reinterpretations of their traditions, their religious underpinnings did not automatically provide all the solutions needed to respond to the impending ecological crisis. However, my study of Swadhyayis, Bishnois, and Bhils presents an alternative environmentalism that is rooted in the dharmic traditions of these communities who have successfully preserved and even developed new ecological resources.

Apffel-Marglin and Parajuli suggest that since the religions tend to separate sacred from profane, they cannot present a comprehensive framework that can inspire the local level collective initiatives that can assume the moral responsibilities for social and ecological justice. I agree that if we view "Hinduism" from the lens of "religion," we might see a divide between sacred and profane. However, most Asian traditions such as Shintoism, Daoism, Confucianism, and Hinduism, lack the theological and organizational foundations of Western religions (Sanford 2007). Alternatively, I suggest that the Indic traditions should be interpreted using the notion of dharma that I present in the next chapter. Dharma can help transcend the dichotomy of sacred and profane because it includes the notion of duty and ethics in addition to religion. Swadhyayis, Bishnois, and Bhils present this dharmic ethos as an alternative to the social ecological framework suggested by Ramachandra Guha (2006). Since duty and ethics are an integral interpretation and component of dharma, dharmic ecology includes the framework of moral ecology that Apffel-Marglin and Parajuli suggest (2000). Apffel-Marglin and Parajuli further present Gandhi as an emblem of moral ecology but fail to note that Gandhi's moral inspiration was deeply rooted in his dharma that transcended the boundaries of Hinduism, Jainism, and even

Christianity. Thus, dharma that includes ethics, morality, duties, and religion should be given serious attention for environmentalism in India and this is the focus of my next chapter.

Chapter 7
Dharma as Religious and Environmental Ethos

In this chapter, I follow McKim Marriott and Gerald Larson in studying Indic traditions through their own categories. Among numerous important terms in Indic traditions, few words can match the ubiquitous presence of the word "dharma" as I will show below. This term has continued to evolve from its origin and today, for millions of Indians, their religiosity and ethics are best expressed in their native vernacular linguistic expressions that are woven around the semantics of "dharma." So much so that, if I was writing this book in any Indian language, I would have been using the words "Hindu Dharma" rather than "Hinduism Religion" to refer to the Indic traditions.

To give an example about the problem of translating "Dharma" as "Religion," in 2002, the then Indian Prime Minister advised Gujarat's Chief Minister to follow his "*Rāja Dharma.*" We cannot translate this into English as "state religion"; rather it means the virtues and duties of political leadership.[1] In his own definition, "A ruler should not make any discrimination between his subjects on the basis of caste, creed, and religion." Indians have been using this term in their daily lives to describe different virtues, duties, and ethics. They also connect their religious lives with ethics using the same term "dharma" interchangeably in different contexts, as I will show below.

Given the overarching presence of dharma in diverse Indian communities of different languages and religiosities, I will explore if it can be adopted to interpret my ethnographic descriptions. In my fieldwork, I noted that Hindus, Jain, and Sikhs of diverse backgrounds use "Dharma" to describe their traditions. People speaking North Indian as well as South Indian languages also use dharma. Just as the English term "Religion" derived from its Latin roots includes its own associated meanings rooted in Western theology (Sanford 2007), "Dharma" derived from its Sanskrit roots brings its unique meanings ranging from virtue, ethics, duty, and even ecology. It can help spread environmentalism because of multivalent significances of dharma in the Indian context.

Although Indian vernacular dictionaries have accepted dharma as the Indic equivalent vernacular term for religious traditions, Monier-Williams Sanskrit-English dictionary gives about 17 meanings including religion, customary observances, law usage, practice, religious or moral merit, virtue, righteousness, duty, justice, piety, morality, and sacrifice (Narayanan 2001). Wilhelm Halbfass

1 I thank Anup Kumar (Cleveland State University) for suggesting this example.

notes that scholarly discussion on dharma is conspicuously absent so far (1988). Halbfass also notes the semantic problems of dharma. Based on a popular Hitopadeśa adage, dharma differentiates humans from animals, apart from the need for food, rest, protection, and sex being the similar traits in both.[2] While this common trait of dharma in all the humans promises an egalitarian message, dharmic emphasis on Varṇāśrama enforces social hierarchy defying egalitarianism (Halbfass 1991). Like Halbfass, Paul Hacker (2006) also observes that the neo-Hindu universalistic interpretation of dharma is different from traditional caste-based interpretations. However, as early as the fifteenth century, long before any contact with Western thought, the Bishnoi guru Jambheśvara had already interpreted dharma as an ethical norm, rather than a caste based duty. Thus, I cannot agree with both Hacker and Halbfass that the universal ethical interpretation of dharma has been influenced by the contact of neo-Hindu thinkers with Western modernism. As Arnold Kunst notes, dharma remains as the intrinsic nature of beings, motivating their conduct[3] (Doniger et al. 1978).

Obviously, scholars have noted the importance of "dharma" in the Indic traditions. Let me now survey the mention of "Dharma" in Indic traditions and texts and its implications for ecology. Dharma is derived from Sanskrit √*dhṛ* meaning to sustain, support, or hold (Holdrege 2004). In the Vedas, *pṛthivim dharmaṇā dhṛtam* signifies dharma as sustainer of the earth. Although the Vedas celebrate the idea of *ṛtam*,[4] cosmic order or rhythm, the term "dharma" also appears in the ṚV, most notably in the Puruṣa-Sūkta 10.90, "*tāni dharmāṇi*." Most scholars agree that Puruṣa-Sūkta is a later addition to the ṚV and hence we can surmise that the idea of ṛtam was being reworded into dharma by the

---

2 Āhāranidrābhayamaithunaṃ ca sāmānyaṃ etat paśubhir narāṇāṃ, dharmo hi teṣāṃ adhiko viśeṣo, dharmeṇa hināḥ paśubhiḥ samānāḥ. Halbfass notes that this popular verse is present only in some editions of Hitopadeśa.

3 Paul Hacker analyzes Bankim Chandra Chatterjee's interpretation of dharma to show how Bankim "confuses" the meanings of dharma in his 1888 book *Dharmatattva*. Hacker also notes a similar trend in the writings of Aurobindo Ghose, Rabindranath Tagore, Mahatma Gandhi, and Vinoba Bhave (Halbfass 1994). He argues that these neo-Hindu attempts to develop a new concept of dharma are based on their borrowing of European positivism and modernism, to find a new norm for ethical and social relationships. However, Arnold Kunst clearly shows that dharma already had similar meanings as early as in the Ṛg Veda (Doniger et al. 1978). I tend to agree with Kunst based on my translation of Bishnoi guru Jambheśvara's discourses that mention dharma in more universalistic ways similar to neo-Hindu thinkers.

4 It also corresponds with the Daoist term Dao, Confucianist term Li, Egyptian term Ma'at, and Avestan term Asha (aša), or Arta, as ethos of several ancient traditions, where people are expected to live in conformity with the cosmic order. See, *Philosophy East and West*, Vol. 22, No. 2, On Dharma and Li (April 1972).

time of Puruṣa-Sūkta.[5] Earlier, yajñas were performed to maintain the ṛtam of the universe. Humans should offer to Agni, the best of the materials, such as *ghee*, grains, and soma. This offering was believed to reach the cosmic gods who would bless them with rewards such as victory, wealth, children, food, and land. In Puruṣa-Sūkta, we see the sacrifice being performed metaphorically, devoid of any graphic details of an actual sacrifice with fire, animals, ghee, or grains. Here, a cosmic man is sacrificed which leads to the creation of the entire cosmos, complete with natural entities such as the sun, the moon, space, the earth, the atmosphere, and the four varṇas. This is then described as the first dharma (Holdrege 2004: 213–48), the cosmic order. Thus, the original notion of exchange among humans and gods in the physical yajñas was already replaced by a cosmic law in the later portions of the ṚV; ṛtam was giving way to dharma even in the ṚV. Thus, the first usage of dharma itself is a reinterpretation of ṛtam in the ṚV. Although both ṛtam and dharma have the sense of cosmic law and order that "sustains" the universe, the mechanism to achieve it shifts significantly from physical sacrifices to metaphorical sacrifice of a cosmic person. This shift also entails the move away from materialistic pursuits to more metaphysical ones, as noted by Joseph Prabhu (cited by Arvind Sharma 2000). Here, we see the first few references of dharma in the Vedas that can be interpreted as cosmic law and order, thus incorporating ecology of the planets, the sun, and other cosmic entities. Dharma is further mentioned in Pūrvamīmāṃsāsūtra of Jaimini, *codanā lakṣaṇo 'rtho dharmaḥ* (JS 1, 1.1). Here, Vedic instructions are the means to understand and practice dharma that in turn is a direct cause of the good *Śreyaskara* and thus becomes desirable (Mohanty 2007). Thus, dharma signifies virtues and righteousness in this context.

Dharma finds different meanings in various other contexts of Hindu, Buddhist, and Jain traditions. In the Jain mantra *Ahiṃsā Paramo Dharma*, non-violence is referred to as the dharma or the supreme virtue. In Jain Saman Suttam 3-5, *Kevali pannattam dhammam saranam pavvajjāmi*, dharma means the teachings of the Kevalins, the supreme teachers of the Jain tradition. In Jain *Sāhu Dhamma*, dharma means conduct and profession of monks (Brown 1954). Śākyamuni Buddha, speaking in a local North Indian dialect, uses the term *Dhamma* as an overarching term to include all his teachings, rules, and laws for monks and lay practitioners. Joining the Buddhist community, *sangha*, one needed to take refuge in this new set of laws by saying *dhammam śaraṇam gachhāmi* (in addition to taking refuge in the Buddha and the Sangha). Thus, the next milestone in the journey of the term dharma finds a new meaning strongly influenced by the Buddha but continuing the underlying significance of law, order, virtues, and

---

5 Arnold Kunst also cites another passage (8.35.13 hymn to the Aśvins) from the Ṛg Veda in which dharma appears in "anthropomorphic representation." Kunst also differentiates dharma from ṛtam and vrata from various Ṛg Vedic verses, ṛtam is the cosmic order, dharma as the social and worldly order, and vrata as the specific duties to maintain the social order (Doniger et al. 1978).

righteousness. Whereas the Vedic dharma rooted in ṛtam was to maintain the universe by performing the roles and duties of social classes and natural entities, the Buddhist Dhamma rooted in the teachings of the Buddha was to be achieved by following him and his words in the community of monks, nuns, and other lay practitioners.

After the death of the Buddha and Mahāvīra, dharma was once again reinterpreted in the texts such as the MBh and the MSm. The Mauryan king Aśoka attempted to establish the rule of dharma in line with Buddhist teachings. According to Patrick Olivelle (2004), it was this emphasis on dharma during the Buddhist period that encouraged later Brahmanical authors to re-emphasize the role and importance of dharma in the texts written afterwards. Thus in the classical period, dharma found itself in a list of four *Puruṣārthas*, the personal-objectives, along with artha (wealth), kāma (desires), and mokṣa (liberation). Thus, the Vedic materialistic ideals were codified under these four objectives restrained by the overarching ideal of dharma, the cosmic order. As J.A.B. van Buitenen has pointed out, "Artha, kāma, and dharma should not be deemed to refer to distinctly different practices – In principle, all three are dharma" (1957). Against the ascetic ideologies advocating nivritti or renunciation as the highest ideal, dharma's definition in the MBh signifies upholding both this-worldly and other-worldly affairs, *Dhāraṇād dharma ity āhur dharmeṇa vidhṛtāḥ prajāh, Yat syād dhāraṇasaṃyuktaṃ sa dharma iti niśchayaḥ* (MBh 12.110.11). Kaṇāda, founder of the Vaisheshika system of philosophy, has given this definition of dharma in his Vaiśeṣika Sūtra, *Yato-bhyudayaniḥsreyasa-siddhiḥ sa dharmaḥ* (That which leads to the attainment of well being and the highest good is dharma) (Mohanty 2007). In the BhG, dharma once again is used in the sense of cosmic law and order when Kṛṣṇa declares that he takes *Avatāra* whenever there is a decline in the dharma, "*Yadā yadā hi dharmasya glānir bhavati bhārata, abhyuthānam adharmasya*." At the same time, Kṛṣṇa inspires Arjuna to follow his *svadharma*, individual duty assigned by his varṇa of kṣatriya. Kṛṣṇa advocates that even imperfect svadharma is better than to do a duty or role (dharma) of a different varṇa. Thus, the BhG underscores the varṇa-roles, as dharma, emphatically against the challenges by Buddhist and Jain traditions.

In the medieval period, after the Islamic invasions, dharma in local languages such as Hindi largely takes the role to distinguish one's religious identity against the Muslims. Suddenly, the term "Hindu" signifies one's dharma, different from "Muslim," as convincingly shown by David Lorenzen (1999). In the late nineteenth century, dharma is used with another traditional term, *sanātana*, to forge a pan-Hindu identity against the reformist ideologies and against the Muslims and Christians in the twentieth century and beyond (Lutgendorf 1991). Thus, we see that dharma has multiple meanings in the Indian context, such as conduct, duty, cosmic law, virtue, and "religion." While "dharma" is often used to translate "religion" in Indian languages, its usage related to virtues, righteousness, and cosmic law and order, can play an important role as argued by Arati Dhand

(2002). From several examples from the MBh, she shows that dharma in the lives of epic characters exemplifies its universal ethical appeal.

In addition to these examples from MBh, Dhand cites examples of dharmic behavior from Rāmāyaṇa. Rāma's foremost concern is with righteousness and he defines dharma as, "Truth, righteousness, strenuous effort, compassion for creatures and kindly words, reverence for Brahmins, gods, and guests" (II.101.30). Dhand concludes that dharma, as exemplified by epic characters, transcends the boundaries of religion and is not limited to matters of soteriology and rituals. Dharma in Indian epics comes closer to the category of ethics, morality, and duties. Following Dhand, I suggest that these ethical interpretations can become the basis for activism in general and for ecology in particular. Ecology comes from the Greek word *oikos*, or home, and works well with the family paradigm of Indian ethics as Dhand shows above.

Obviously, Indic traditions based on the concept of dharma incorporate duties, virtues, ethics, and spirituality simultaneously. Following Dhand (2002) and Narayanan (2001), I would like to note that the influence of epics such as the Rāmāyaṇa, the MBh, and the Purāṇas is widespread in India, both in the rural parts of North India and South India. As Dhand shows, the characters of these epics, especially the heroes such as Rāma and Yudhiṣṭhira, who sacrifice their personal interests to serve and protect the ideals of dharma, exemplify the idea of dharma. Moreover, the sojourn of the epic heroes such as Rāma and Yudhiṣṭhira into the forest is called "the seed of dharma" in folklore because the forest acts as a place of testing and their period of forest exile is seen as a kind of initiation prior to their assumption of rule. Anne Feldhaus notes from several Sanskrit sources that the forest is associated with dharma, the social and moral order that is supposed to rule life in the village, the city, and the kingdom (Feldhaus 1995: 102).

In addition, the MBh defines dharma, as I noted above, as one that *sustains* both the personal order and the cosmic order. Indians, in their daily lives, use dharma interchangeably to describe their ethos as it relates to their religion and natural order. Especially for the rural Indians the distinction between the religious ethos and the ecological order is negligible since they describe them with the common term dharma or *dharam*. Several scholars have noted this trend in Indians. Ann Gold's observations from her fieldwork in Rajasthan are especially helpful. She describes the villagers who relate their moral actions with ecological outcomes (Gold and Gujar 2002). Frederick Smith records similar trends in the ethnosociology of Marriott and Inden (2006: 586). Smith also cites Arjun Appadurai, "South Asians do not separate the moral from natural order, act from actor, person from collectivity, and everyday life from the realm of the transcendent." Smith concludes, "The distinction between mind and body, humanity and nature, essence, idea, quality, and deity, would be (largely) one of degree rather than of kind."

Today, while the Western scholars have already started developing ethical theories to respond to environmental challenges, non-Western cultures in general and Indians in particular have not yet begun in this direction, as noted by Austin Creel.[6] I now respond to some of the questions raised by Creel regarding the validity of dharma for Hindu ethics in the modern world.

Creel notes that the Hindu philosophers were oriented toward mokṣa rather than dharma, and these were not integrally related. However, I can only partially agree with this. This assertion seems to be based on interpreting the diverse set of Indic traditions with a renunciatory lens as was done by Orientalists, as shown by Haberman and others. I would argue that Indian epic heroes had successfully transcended the dichotomy of dharma and mokṣa. Dharma served as the way of life for them to attain the mokṣa. This "virtue ethics" has served as a role model for Indians for several millennia (Matilal 2002). From my research with the Swadhyayis and the Bishnois, I found that their inspirations were the Hindu epic heroes and their gurus whom they see as practitioners of dharma to attain the mokṣa. In several of the discourses of Athavale, I found him exhorting his followers to follow the ideal of Arjuna, the warrior of the Mahābhārata who preferred the path of the *pravṛtti* (action) instead of *nivṛtti* (renunciation). Athavale repeatedly stressed that only actions done with a devotional motive can be considered dharmic actions leading to mokṣa. Thus, he correlated the motive of the action with the potential for mokṣa. Bishnois also told me that Jambheśvara was a preacher of the Karmayoga. Thus, the first step to develop an environmental ethic based on my case studies would be to develop an ethical framework based on dharma and karma that is integrated with mokṣa, not in opposition to it, as is claimed by Creel.

Creel further raises doubts about the monism, a characteristic of mokṣa, which seems to be world-denying and hence antagonistic to the dharma which is world-affirming. Here, he seems to be overemphasizing the monism of Śaṅkara and ignoring the philosophies of Rāmānuja and Vallabha who sought to integrate the householder values related to dharma with monistic values related to mokṣa (Ingalls 1957). Although Athavale paid lip service to Śaṅkara,[7] his interpretation of monism is closer to the *Viśiṣṭādvaita* of Rāmānuja. While the monism of Śaṅkara led to a sort of cosmological dualism distancing orthopraxis into the category of unreal and insignificant, theistic Vedantic thinkers such as Rāmānuja and Vallabha sought to rectify this by according the world an ontological continuity

---

[6] Austin Creel notes that Hindus have not taken a systematic approach to develop ethics as a discipline of study, unlike the Western thinkers (1977).

[7] Athavale called Śaṅkara the greatest Indic philosopher especially as known to the Western world due to the missionary zeal of Swami Vivekananda. Nicholas Gier (2007) makes an excellent remark about such lip service by Indians, "Many Indian philosophers, after lecturing on Advaita Vedanta, go home and make offerings to Ganeśa. Just as no European ever worshipped Aristotle's unmoved mover, no Hindu has ever bowed before nirguṇa Brahman."

with Brahman. Athavale repeatedly emphasized that the world is divine because the Almighty resides in it and hence dharmic duties are not in opposition to mokṣa, they are integral to one another (Rukmani 1999). According to Nicholas Gier (1995), Rāmānuja's panentheism is a much better philosophical model for environmentalism. There is a significant contrast between seeing the cosmos as a dream-like appearance (Śaṅkara) and the cosmos as the very body of the Godhead (Rāmānuja). The ultimate effect is that absolute monism desacralizes the universe, while panentheism resacralizes it (Bartley 2002).

The second concern raised by Creel is that dharma is correlated with a static society with little possibility for change. With India changing at a rapid pace in last few decades, it may pose special obstacles for the dharma concept that was centered on stability. However, both the Swadhyayis and the Bishnois seem to have no problem in correlating their dharma in modern times. The Swadhyaya movement, especially, has its origin and development precisely in the period of modernization and industrialization launched after Indian independence. Instead of rejecting industrial revolution, Athavale sought to make use of the modern technology to propagate his ideology based on dharma. Similarly, Bishnois made use of new political developments to join the democratic process. Instead of finding their tradition at odds with modernity, we find Bishnois at key positions in the state and local government handling the departments of foresty and animal protection (Fisher 1997). Thus, the advent of modernity has facilitated the Swadhyayis and the Bishnois to practice their dharma more effectively instead of posing an obstacle.

The next problem that Creel notes is the lack of a central point of authority in Hinduism for expressing a position about dharma and other issues.[8] Although one can easily agree with this observation of Creel, the absence of a central authority in Hinduism has led to a healthy flexibility and tolerance of different customs and virtues instead of authoritarian laws imposed from above as shown by Whitney Sanford with regard to Krishna worship (2005). In my case studies, Athavale and Jambheśvara developed their own ethical theories for their distinct communities and sought to validate it within the broad framework of dharma by supporting their teachings with quotations from Hindu epics such as the Rāmāyaṇa and the MBh. Their teachings worked effectively for thousands of their followers, even in the current times, suggesting that this ethical framework has worked on the ground even though both cannot be called philosophers of ethics in the Western sense.

The next concern of Creel is that the neo-Hindu interpretation's emphasis on social equality is in sharp contrast to the traditional dharmic view of recognizing individual roles based on individual castes and varnas. He argues that dharma as a view of innate law would require attention to individual characteristics in defining roles; otherwise, the traditional connotation of dharma is being

---

8 One can add several new ethical issues on which there is no central "Hindu" position, e.g. stem cell research.

supplanted by a view of ethics based on the presence of Brahman in everybody, repudiating the theory as well as the details of dharma. Here again he sees a caste system based on traditional dharma to be in opposition with the spiritual equality based on the monism of Vedanta. I would like to respond to this by taking note that both the Swadhyayis and the Bishnois have emerged as a new movement in different periods in Indian history. The founders of both movements sought to integrate people from different castes in a new unified community based on the spiritual equality in them.[9] Although Creel's concerns seem philosophically valid, the evidence from my fieldwork suggests that at least in the ethical framework of Jambheśvara and Athavale, dharma is not in opposition to mokṣa. Alternatively, I agree with McKim Marriott who noted that dharma, comprising of concepts of attributes (guṇa), power (śakti), and action (karma), is felt by Hindu actors to be directly involved in administration, conflict, and leadership.

Several authors such as T.S. Rukmani, George James, Vasudha Narayanan, and O.P. Dwivedi have shown theoretical and textual references from Indian texts (Chapple and Tucker 2000). It is also apparent that these textual references have also been the part of the worldview or cosmology of Indians for the last several millennia. They follow the texts by revering natural resources. Nature worship is thus to be found not only in various texts but is also evident from the practices of Indians. Thus, there seems a reflection of texts in the religious practices of Indians. However, this textual and cosmological semblance is hardly found in the behavior of the people *outside* their religious sphere. As Tomalin, Nelson, and others have noted, the dharmic teachings from the epics does not automatically inspire Indians to become environmentalists. On the other hand, I found the Swadhyaya movement and the Bishnois who do seem to be connecting their texts with their rituals and their ethos. Their relationship with nature not only includes revering it, but it also seems to inspire them to restore, protect, and conserve it.

This brings me to summarize the theoretical argument of sociologist Grant McCall (1982). He reviewed the work of several anthropologists and found that they describe the practice of dharma in Indian rural societies but do not mention the word dharma in their interpretive analysis. On the other hand, other works mention the word dharma but do not describe how dharma continues to play a major role in the lives of Indian villages. In an attempt to fill this lacuna, he proposed that the lives of Indian rural societies should be interpreted using dharma rather than the Western conceptual framework based on religion. He argued that dharma helps explain the village life better than other more

[9] R.J. Fisher (1997) notes that Bishnois have the traces of their sub-castes but they largely identify themselves by belonging to the Bishnoi caste. Like several other movements, the early egalitarian zeal of Jambheśvara against the caste system yielded to a new caste of Bishnoi. Similarly, Parel (2006: 13) notes that Gandhi sought to create a caste-free egalitarian society in his Sabarmati Ashram in Ahmadabad.

commonly deployed concepts such as the caste system. Like other societies, order is the desired goal of the Indian society also and dharma best describes this order of the Indian society, just as *jen* and *li* describe that for the Chinese, and *logos*, *ethos*,[10] and *eusebeia*[11] for the Greek society. Dharma is the indigenous organizing principle in terms of the Indian belief system. It is like a numinous power for an Indian community responsible for the maintenance and conduct of both the social order and the world in general.

Thus, morality and natural phenomena are connected and interdependent. This organizing principle also matches with Edgerton's notion of the "dominant idea" in a people's culture (1942). Dharma occurs, in identical or semantically equivalent forms, frequently in Indian texts. Both the authors of these texts and the lay Indian society regard it as an important notion for its bearing on human life and conduct. My research suggests that dharma appears with a high degree of frequency in the texts and daily conversation of Indians as an explanatory principle and that the people's behavior conforms to their professed beliefs. Therefore, agreeing with McCall, I suggest that dharma can be elevated from a folk or Brahminical notion to an analytical level, especially as it pertains to both the religious and ecological "attitudes" of Indians (Potter 1991). Like any other society, Indians of different backgrounds such as different languages, castes, and regions, subscribe to a concept of order as the most desirable end, with each group (and each person in that group) holding a unique understanding of what constituted that overall orientation. Thus, individuals interpret and apply dharma in their own situations freely even though there are overarching generic laws and norms laid down by the Indic traditions based on dharma.

Ariel Glucklich (1994) has suggested that the most fruitful approach to understand dharma is to set aside the quest for conceptual framework and theoretical formulations and to adopt instead a phenomenology of dharma based on a "somatic hermeneutic" that explores embodied experiences of dharma in specific spatial and temporal contexts. Glucklich convincingly employs Wolfgang Kohler's Gestalt psychology to offer a more satisfying psychological analysis as to how Indian rituals, such as river bathing (immersion in water), result in a psychosomatic purification that produces a new state of consciousness. He cites Hindu bathing as having power and meaning, not through sociological (structural or functional) or conceptual a priori systems, but through a symbolic process in which embodied sensory experiences play a dominant role in evoking a new and transforming (purifying) state of consciousness. Glucklich recognizes and tries to overcome the Cartesian conditioning that focuses on a mental conceptual analysis but ignores the key body side in the Indic experience of

[10] Nicholas Gier notes the resemblance in Chinese li, Greek ethos, and Indic dharma (2005).

[11] An inscription of the Indian emperor Ashoka from the year 258 BC was found in Kandahar in Afghanistan that used the Greek rendering eusebia of the Indic word dharma (Hacker 2006).

dharma. Glucklich maintains that the body, mind, and natural environment must be studied as a gestalt. He argues that focusing on the images of embodied experience, rather than on noumenal concepts, helps to evoke the temporal resonance of the text and bring its dharmic experience "to life." In this way, the Cartesian dualism of mind and body is transcended via sensitivity to the powerful environment, which evokes a different mind and consciousness, for example, for the early-morning bather in the Ganges. Glucklich calls this new resulting mind "the embodied imagination where perceptions, self-perception, and symbolic ideas resonate together."

Glucklich's excellent phenomenological study of dharma seeks to correct previous approaches that have fallen into the Cartesian trap of seeking to understand Hindu dharma through mental categories only. Instead of superimposing the Western Cartesian mindset on Hindu dharma, as many previous studies have done, Glucklich examines dharma as a body-mind-environment gestalt. Thus, considerations of Hindu dharma must extend from mental textual constructs to daily experiences by the body in its immediate cosmic environment where the world is imagined as a transparent unity. As the stream of sensory experience is constantly flowing, dharma only has the appearance of permanence. While the dharma texts show that dharma boundaries are fixed and absolute, the flow of bodily experience, upon which such boundary conditions are superimposed, is constantly changing. The ambiguity that results is often better reflected in the myths of the epics and purāṇas than in the dharma texts themselves. Thus, Hindu dharma manifestations at the level of bodily perception (house walls, field boundaries, rivers, etc.) are important for the study of Indian culture.

I have tried to explore such patterns in my case studies. By participating in different activities related to ecology, the practitioners of traditional communities such as the Swadhyayis and the Bishnois not only undergo somatic experiences but also these experiences help them to "relive" the lives of Vedic sages and other mythical figures such as Arjuna. This is the embodied imagination or the "ecological mind" where perceptions, self-perception, and symbolic ideas resonate together. This is the level at which dharma means something to Hindus before it has acquired its extremely diverse lexical meanings and social functions. It connects the practitioners with the experiences of their gurus and their natural surroundings.

Weightman and Pandey, two Hindi lecturers in London, analyzed hundreds of Hindi sentences and found that the word dharma in everyday language of North Indians chiefly signifies three things: religion, duty, and intrinsic property (Doniger et al. 1978). In my fieldwork, in addition to these meanings, I also saw that the third meaning of intrinsic property or attribute of an inanimate object was also assigned to the cosmic entities such as the sun, the moon, and the earth. For instance, I heard from my informants that the dharma of the sun (and the fire) is to burn, the dharma of the earth is to revolve around the sun, and so forth.

In the foregoing analysis, I tried to present the sociological arguments by McCall, psychological argument by Glucklich, and linguistic argument by

Weightman and Pandey to show the relevance of dharma at different levels in Indian society. In addition, my survey about the evolution of the term dharma suggests that dharma can function as a bridge between the ecological notions and environmental ethics of local communities of India and the scientific ecological message related to the planet earth. I suggest that the word dharma can be effectively used to translate the scientific ecological awareness to reach out to the local communities of India.

Pramod Parajuli (2001) presents a similar model of intersecting human, natural, and supernatural domains and argues that this intersecting area is the actual world of practice. This is where humans are engaged in deriving their livelihoods by taking care, and reshaping their material culture. Based on my analysis of dharma in this chapter, it seems suitable for this intersecting area that Parajuli is suggesting as an intersection of human, natural, and supernatural worlds. In the human world, dharma refers to social duty. In the natural world, it refers to the intrinsic property of ecological entities. In the supernatural world, it refers to the matters related to religion. Following Dhand, Narayanan, Gold, and others, I see dharma as an ideal bridge signifying ethics, virtues, duties, and cosmic order. In a seminal way, Glucklich also shows us that dharma indeed enables Indians to transcend the Cartesian divides of body, mind, and natural world. McCall, Marriott, and Fitzgerald have already suggested that dharma can be an emic category to study and analyze the Indic world that frequently transgresses the world of religion, environmental ethics, and human social order, as is evident from my case studies of Bishnois, Swadhyayis, and Bhils and from the linguistic survey by Weightman and Pandey. Narayanan (2001) also records that Hindus use the word dharma to mean both religion and ethics. With this background of dharma, I now move on to present my conclusions involving my case studies in the next chapter.

Chapter 8
Conclusions

Before I compare my case studies of the Swadhyaya movement, the Bishnois, and the Bhils, I first go back to the two different kinds of environmentalism based on the devotional and the ascetic models that I reviewed earlier. All my case studies involved Indian rural communities spread in the villages of Western India. Based on my observations of them, I found that their practices derive from the devotional model that is different from the ascetic model. To be sure, the devotional Hindus do not reject ascetics. They continue to attend discourses by ascetics and pay their respect to them but when it comes to their own personal practices, it is the daily rituals, pūjā, at home and at temples that heavily dominate over any austere practice by a layperson (Madan 1996). For example, the Bishnois gladly visit the pilgrimage sites where their guru attained enlightenment, but on returning home, they would indulge not in similar austere practice, which would lead toward an ascetic enlightenment, but they would come back to their routine daily rituals, such as *sandhyā* and *ārati*. Lay Hindus keep fasts in a milder form than its ascetic versions, which are much more austere. Compared to lay Hindus, lay Jains' fasts are much more austere. While lay Hindus would eat fruits and vegetables in their fasts, lay Jains avoid water and all kinds of food just as ascetics (Jaini 1979: 157–85).

This last point brings us to another interesting dimension of environmentalism inspired by Indic traditions. Our two models of devotional and ascetic actually lead us into a dichotomy of the Hindu traditions and the Jain traditions. As we saw, the majority of Hindu practitioners follow devotion in their daily rituals, and extending our discussion to Jain laity, we find that Jain lay practitioners come much closer to the austere practices of ascetics. Jain role models are their Tīrthaṅkaras who had renounced all their belongings, including their clothing, to perform the toughest austerities possible. Even the temple-going Jains know that the Jain ideal is to renounce householder life and to follow the path of their role models such as Mahāvīra, other Tīrthaṅkaras, and the contemporary monks and nuns (Vallely 2002b). The Jain ideal is to attain mokṣa by renouncing worldly life, whereas for most Hindus, especially the followers of Vallabha and Rāmānuja, the ideal is to become the perfect devotee or attain mokṣa by practicing their routine householder lives.

Naturally, scholars of Jain environmental ethics, such as Christopher Chapple, have advocated the ascetic model for environmental ethics in their writings, while scholars of Hindu environmental ethics, such as Vasudha Narayanan and David Haberman, have advocated the devotional model. I think, overall, we can combine these two models to help restore and replenish the disastrous ecology of India and beyond. While the ascetic model can help reduce the over-exploitation of natural resources by limiting one's desire for more luxuries, the devotional model can help restore natural resources to their original beauty and harmony. The ascetic model will need to be prescribed for people of higher classes and developed societies, those who continue to plunder the planet for their extravagant consumption. While describing American society, Diana Eck posted this on the *Washington Post* blog (December 14, 2006), echoing Gandhi's prophetic words, "The earth has enough for one's need but not for one's greed."

> Is it a moral good to consume far more than our share of non-renewable energy resources, creating for us a standard of living that does not know the meaning of the word 'enough' and that acquiesces in a world of unconscionable economic disparities?

On the other hand, the devotional model may be suitable for both the higher and the lower classes. As Haberman demonstrates, most Hindu environmentalists are inspired by their devotion toward the rivers. Swadhyaya prayogs also show the similar devotion for trees, water, earth, and cattle. Similarly, Bishnois and Bhils practice their religiosity chiefly by devotional practices of rituals, temple visits, and pilgrimage to their respective sacred places. While the Swadhyayis and the Bishnois are mainly farmers and vegetarians, Bhils are mainly hunter-gatherers and meat-eaters. While Bishnois today are seen to be political and social rivals of the other powerful groups such as Rajputs and Jats, Bhils are still seen as a tribal community on the fringes of the mainstream society. Despite these obvious differences in them, they have tried to save and protect trees in their villages based on their devotion for natural resources. Although the founder of the Bishnoi sect, Jambheśvara, led a life as a celibate ascetic and performed austerities in the hot desert of Western Rajasthan, his followers are mostly householders leading a life of daily rituals. On the other hand, Athavale, founder of Swadhyaya, led a householder life and his followers are householders. Unlike Bishnois and Bhils, Swadhyayis neither are on the fringes of society nor socio-politically as monolithically organized as a Bishnoi "caste-group." Christopher Chapple recognized the dichotomy of devotional or world-affirming model and ascetic or world-denying model of Indic traditions and sought to see an underlying common theme in both (1998b).

Following Chapple's attempt to transcend the dichotomy of devotional and ascetic models, I want to extend the notion of dharma further by combining religion, ethics, and ecology. One of the fundamental problems in studying or researching Indic traditions is the search for Western categories of knowledge within them. Scholars have long wrestled with various Western categories such as religion, ethics, theology, and history and their Indic equivalents. Gerald Larson spoke about the need to apply Indic categories of knowledge to the study of India instead of looking for Western categories (2004: 1003–20). McKim Marriott's ethnosociology of India is rooted on the same philosophical problem (1990: 1).

Elsewhere Marriott notes that Western history has separated various domains of knowledge such as religion, psychology, sociology, anthropology, and (if I may add) ethics and ecology, but the scholars should not assume that the non-Western cultures would also wish to divide them. Following Marriott, I propose not to see environmentalism, ethics, or theology as separate categories in Indic traditions, and suggest that ethics, ecology, and theology are all intertwined in Indic traditions as exemplified by various texts, recent movements, and my ethnographic encounters. I am positing this intertwined relationship in a "dharmic" framework rather than a "religious" one. Vijaya Nagarajan extends Karl Polanyi's understanding of embedded economies to that of embedded ecologies (1998: 165–95).

I would like to extend this "embedded" notion further by combining ethics, ecology, and theology with an overarching term "dharma" in which they are intertwined due to its varied interpretations. In January 2005, after a gap of 10 years, I visited the marble mines near Udaipur, a major tourist destination in Southern Rajasthan. In 1995, when I had last visited there, it did not suffer from scarcity of water. However, after 10 years, the green valleys of the Aravallis seemed dry[1] and so I asked about the rainfall situation. My taxi-driver told me this, "Earlier we never faced lack of rainfall in this area but since the advent of Marwaris (business community from Western Rajasthan) for the lucrative business of marble (by strip mining),[2] God is punishing us for their greed." This remark by a local taxi-driver matches similar comments Ann Gold received in her ethnographic accounts of residents of Ghatiyali (1998: 165–95) in which deforestation was associated with numerous alterations in human relationships, both political and interpersonal. Gold's observations also resemble Peter Gaeffke's argument that "justification of God" is not the same as "God's justice." Thus, karmic law of action and consequence is not the same as "judgment by God" in the Abrahmic religious sense but it is a "mechanical justification" resulting from one's action (1985).

This relationship of human actions with ecological ramifications was also famously noted by Mahatma Gandhi after a severe earthquake in Bihar in 1934, "A man like me cannot but believe that this earthquake in Bihar is a divine chastisement sent by God for our sins." What these remarks by Gandhi and by natives of Rajasthan suggest is a deep sense of interconnectedness of human behavior and natural phenomena. This interconnectedness can also be seen in Indic texts. The twentieth-century Indian leaders such as Gandhi and Athavale inspired their followers to build and nurture this relationship.

---

1 An Australian anthropologist R.J. Fisher returned to Western Rajasthan in 1994, after his earlier fieldwork there in 1987, and contrary to my observation, he in fact noticed an increase in vegetation and he suggests that desertification and soil degradation is not as prevalent as it is sometimes assumed (Fisher 1997: 221).

2 The Department of Mines and Geology, Government of Rajasthan has laws for eco-friendly mining but their implementation remains questionable (Sebastian 2006).

Perhaps it is because of this intertwined relationship of spirituality, ethics, and environmentalism that scholars such as John Cort, Paul Dundas (Chapple 2002), Peter Flügel (2005), and Patricia Mumme (Nelson 1998) have noted that environmental ethics does not exist in the Hindu or the Jain traditions. However, based on the dharmic ethos in these texts as argued by Dhand and others, ethics and environmentalism are intermingled even though they do not have evolved theories. Moreover, my observations of the three communities also match this textual dharmic ethos. The Swadhyayis, the Bishnois, and the Bhils also express that human behaviors are irrevocably interwoven with environmental conditions; the deterioration of one implies and involves the other. While the "transactional model" of Marriott and Inden (1981) suggests a horizontal mutual relationship among different castes and classes of "dividuals" of Hindu society, I suggest that this transaction includes ecological resources. This "attitude," as Potter refers to the four Hindu *puruṣārthas* (1991), also transcends class, caste, or gender divide suggesting that it is not just a viewpoint of one group of society. Thus, I would like to situate my observations with those of David Haberman, Ann Gold, Vasudha Narayanan, and George James and with the theoretical framework of the "Man-Nature-Spirit complex" developed by Mann and Vidyarthi. The intertwined relationship of humans with ecology and divinity seems to be the hallmark of Indian society.

This intersecting relationship is what I tried to describe, making use of the term "dharma," i.e., a combination of ethics (in the human world), ecology (in natural world), and theology (in the spiritual world). Agreeing with Apffel-Marglin and Parajuli (2000), I suggest that the term "religion" splits itself from the "secular" world, especially after the Western Enlightenment. However, the term "dharma" is based on "Vedic monism," a holistic idea that means virtue (in the human world), cosmic order (in the ecological world), and rituals (in the spiritual world), as I noted in the previous chapter. Weightman and Pandey also found that Indians make use of this term interchangeably to mean these three different ideas in different circumstances in their vernacular languages.

While it is true that environmentalism as a category does not exist in Indic traditions, it is equally true that the dharmic Indic traditions have helped sustain the Indian ecology for several millennia by inspiring Indians to limit their needs. One of my students, Timothy J. Hulme, Jr., put it succinctly. What sets humans above beasts is their ability to cease or control animal urges. Few animals can control eating, refuse mating, or censure diet. This makes ascetics, fasting, celibacy, and vegetarianism fascinating to him (though not in his own practice) and it is India that is the first place all these things occurred. This makes him think India may be the birthplace of humanity.

Although asceticism, fasting, and celibacy are practiced only by a minority of Indians, the main diet of the majority of Indians largely consists of rice, wheat, pulses, and vegetables. Even those who are classified as "non-vegetarians" depend largely on vegetarian food as the chief components of their diet while eggs, meat, and fish are consumed occasionally. In 2002, India's meat consumption was 5,456,264 metric tons, much less than other major meat-consuming regions.[3]

This shows that even after the advent of modernity and globalization Indians have successfully preserved their vegetarian habits that were laid down by their dharmic traditions several millennia ago. Interestingly, meat-eating is now linked to global warming. In a groundbreaking 2006 report, the United Nations said that raising animals for food generates more greenhouse gases than all the cars and trucks in the world combined. Senior UN Food and Agriculture Organization official Henning Steinfeld reported that the meat industry is "one of the most significant contributors to today's most serious environmental problems." On the one hand, we find a long tradition of avoiding the meat in Indian dietary habits and, on the other hand, the latest reports from the UN declare that meat-eating is one of the main reasons for global warming. Even after Western media[4] reported about the connection of meat-eating with global warming, leading environmentalists such as Al Gore, who got the Nobel Prize for his work in this regard, failed to take any notice of meat consumption in the food habits of Western society.[5]

Even such clear evidences have so far been ignored by the Western society in general and the environmentalists, such as Gore, in particular. Thomas Friedman, a leading *New York Times* columnist, noted this and even rejected any changes needed in the Western lifestyle even while demanding the "greener" initiatives from the US government (*New York Times*, April 15, 2007). This Western dichotomy between expecting the "environmentalist" initiatives from the governments and businesses *without* changing personal lifestyles was the subject of the conclusion of Ramachandra Guha's book with an appropriate title, "How much should a person consume?" Guha observes that the Western society consists of 20 percent of the world but consumes about 80 percent of the production of the world. The rest of the world consisting of the 80 percent of the world population consumes only about 20 percent of the production of the world. Guha agrees with conservationist Ashish Kothari and criticizes the "hypocrisy" of the developed world (2006: 149): "It is, *the allegedly civilized,* who have decimated forests and the wildlife that previously sustained both tiger and tribal. With rifles and quest for trophies, [they] first hunted wild species to extinction; now [they] disguise [themselves] as conservationists and complain

---

3 In the same year, Chinese consumption was 67,798,988, European consumption was 53,996,792, North American consumption was 39,716,290, South American consumption was 24,873,257, Middle Eastern and North African consumption was 9,524,500 and Central American and Caribbean consumption was 8,179,695 metric tons. See: Food and Agriculture Organization of the United Nations (FAO), FAOSTAT on-line statistical service (FAO, Rome, 2004).

4 Such as CNN, Fox News, *New York Times*, *Time* magazine, *Guardian* (UK), and *Telegraph* (UK).

5 http://blog.peta.org/archives/al_gore (viewed on December 20, 2007).

that adivasis are getting in the way. The real 'population problem' is in America, where the birth of one child has the same impact on the global environment as the birth of about seventy Indonesian children. Worse, the birth of an American dog or cat was the ecological equivalent of the birth of a dozen Bangladeshi children."

What is even more striking is that due to the dharmic traditions inspired and founded by gurus and sages such as the Buddha and Mahāvīra, Indian society had successfully moved away from animal sacrifices and killings prevalent in the Vedic era to lifestyles largely based on vegetarianism. Ironically, scholars seem to have largely ignored vegetarianism as one of the most important dharmic lessons inspired by Indic tradition that can greatly help reduce global warming. Incidentally, both Bishnois and Swadhyayis are vegetarians and even Bhils have turned to vegetarianism especially after the Bhagat movements' influences on them as I have shown above. Out of several such lifestyle changes that were inspired by the dharmic teachings of the Buddha and others, I have just shown one here. We can similarly note others, such as *Aparigraha* (non-accumulation), which have continued to be an "obstacle" against the consumerist revolution in India. Only in the 1990s, finally, did India start embracing the Western capitalist model of economy and now market forces are fast transcending the proverbial "Hindu rate of economic growth." Until this Western market invasion, the so-called Hindu rate of growth might have been both the result and the reason for limited Indian spending on consumer goods (Gold 2001).[6]

It is true that despite the presence of ecological reverence and appreciation in Asian traditions, they could not stop the ecological devastation in India, China, Japan, and much of Asia in last couple of centuries. Eugene C. Hargrove (who co-founded the first American ecological journal, *Environmental Ethics*, in the 1970s) responded to this common scholarly critique (1989: xix–xxi). He agreed with the criticism that Asian traditions could not stop the ecological devastation even before the Western entry into Asia but it was not because Asian traditions lacked the ecological resources. He conclude that precolonial environmental degradation in Asia, occurring gradually over centuries, was the result of a lack of scientific knowledge of ecological relationships – rather than the pragmatic inefficacy of Eastern environmental attitudes and values.

Introducing the same volume, the editors called the emerging discipline of environmental ethics "anti-applied ethics." They noted that other kinds of ethics such as biomedical and professional ethics are called applied ethics because they merely apply the traditional Western philosophies to the emerging fields of biomedical and other professions (Callicott and Ames 1989). Extending this observation further, I suggest that environmental ethics for Western philosophical traditions may be "anti-applied ethics" but, for Asian traditions, it

---

[6] See Gold (2001) on how consumption is severely constrained and morally limited by ideals of self-restraint in Hindu traditions – fasting, eating only what is appropriate and so forth.

is truly applied ethics. As Hargrove noted, Asian ideals and moral principles can be valuable resources for ecology if applied in the practical world. This is what I suggested in my discussion of dharma in the previous chapter. If this dharmic ecology can be applied to wider society beyond the Swadhyayis and the Bishnois, India's environmental problems can be solved.

Although I agree that we should not equate modern environmentalism with ancient wisdom traditions' reverence for nature, I would now attempt to discuss Hindu nature worship from different theoretical angles. As Rich Freeman noted in Kerala, the villagers worshipped the sacred groves but did not practice environmentalism. The same can be applied to overall Indian and other Asian traditions. Right from the days of the Indus Valley Civilization and the Vedic Era, we see much archaeological and textual evidence of nature worship in Indian society, but we do not find the same environmentalist activism or awareness that we have seen in last few decades around the world. The question arises, why Indians worshipped nature if they did not realize its importance for their survival. Why did they revere Ganga, Himalaya, Tulasi, and their cows if they had no concern for biodiversity and environmental pollution? We already saw several examples of contemporary Christian organizations adapting to "become green." We also saw several examples of neo-Hindu movements and organizations also including several environmental practices to "become green."

However, we may never be able to know the exact motivation and intention of the early Indians or people of other folk traditions who worshipped nature around them (Nugteren 2005: 1). Any attempt to decipher the original intention of early Indians will be merely a hermeneutic exercise. Here, I found the approach of Frederick Smith, building upon Catherine Bell, to be more appropriate (2006: 152). Smith utilizes Bell's term "traditionalization" to argue that the widespread practice of possession in Indian society must have been gradually incorporated into the Brahminical texts and practices based on continuous interactions between the folk and classical traditions. Following his argument, we can assume that diverse Indian folk communities and tribes must have been worshipping diverse natural powers such as the sun, the earth, the rivers, the mountains, different animals, birds, and trees and so on. With the gradual interactions of Aryans and non-Aryans, much of the rituals and myths of diverse communities must have been *traditionalized* into Brahminical texts and thus several gods and goddesses assumed pan-Indian popularity. These deities either represented natural resources themselves or incorporated animals and birds as their companions. This "traditionalized" nature worship lacks the formalized or institutionalized philosophy that has rendered it largely invisible to scholars looking for environmental ethics in India. Furthermore, Smith shows that although possession was traditionalized, it was never institutionalized (2006: 152).

I find this equally applicable to nature worship in Indian rituals. While the omnipresence of divinity was formulated into Advaita and other philosophies, the outward manifestations of this underlying divinity remained a traditionalized form in myriad symbols and deities rather than a philosophized idea into the Brahminical texts. Ramachandra Guha cites Nobel Laureate George Akerlof who remarked of his fellow economists that if you showed them something that worked in practice, they would not be satisfied unless it was also seen to work in theory (2006: 1). Just because Indic traditions lack the Western notion of a *theory* of environmental ethics, their *practice* of environmentalism embedded in their daily lives should not be ignored, as also suggested by Purushottama Bilimoria et al. (2007: 17).

Following Bina Gupta (2006), I would like to suggest that dharma is comprised of rules that have been handed down over generations for social cohesion. Instead of an "absolute" set of such rules, dharma has provided a series of markers to Indian society. However, the decision has had to be made as to how to apply these rules in practice. Dharma has rarely imposed a sort of discipline that exacts an obedience, which determines every significant decision that one makes. There has been considerable latitude on this count, and the question of how to reconcile the different dharmic rules has arisen frequently. The rules of dharma are not derivable from a single principle, or even from a single set of principles; these rules are learned from epic stories, the widely used source of knowledge in the traditional epistemology, and can vary in individual circumstances depending on the context. Two justifications are traditionally provided for this – first, that *śabda* (word) is an infallible source of knowledge, and second, that self-realization is the highest goal of human pursuit. It must be noted that *śabda pramāṇa* applies only to *śruti*, and not to the *dharma śāstras*. Thus, there is no authoritative support for the caste duties, for anything taught by the *smṛtis* is, in principle, fallible and revisable. In Hindu ethics, the term "dharma" is used for both "duty" and "virtue." The concept of "dharma," like the German "*Recht*," covers a large spectrum of different but connected meanings. It encompasses within its fold a theory of ethical rules, a theory of virtue, a social ethics, and an account of the Kantian notion of duty for duty's sake, leading to the goal of mokṣa. There are two components of dharma, the subjective and the objective. The subjective dharma is concerned with inner purification, purification of the mind, inner discipline; the objective dharma is concerned with duties, including universal or common duties and those duties that depend on a person's particular position in society and stage of life.

In her book on bioethics, Swasti Bhattacharyya (2006) shows how Hindu narratives can become a source of ethics. Her study of the MBh demonstrates a commitment to a variety of beliefs that she lists as: (1) the centrality of society; (2) the belief in an underlying unity of all life; (3) the responsibilities and flexibility of dharma; (4) the multivalent nature of Hinduism; (5) a theory of karma; and (6) the teachings of ahiṃsā. The six elements that she applies for bioethics can also be applied for environmental ethics.

The first element regarding the centrality of society leads to an ideal where individuals are encouraged to act in a manner that supports and works toward the greater good of society. The first element in turn connects with the second element regarding an underlying unity of all life, the world, and all that exists. All of society is part of a cosmic body. In order for it to operate smoothly and to ensure cosmic order, each portion of society needs to fulfill its function; each individual needs to perform one's duty. Dharma is the third element that provides guidelines for how the individual and society are to function; these guidelines are not abstract universal principles. Rather, each individual is encouraged to particularize and contextualize dharma to individual situations.[7] This flexibility is underscored in the fourth element regarding the multivalent nature of Hindu traditions. This flexibility and freedom of an individual's choice regarding one's dharma is linked to the next element about the theory of karma. Instead of criticizing or punishing individuals for making their individual choices, which sometimes may not comply with dharma, most Indian traditions demonstrate that individuals always reap the consequences and benefits of their decisions and actions (Bhattacharyya 2006: 73).

Thus, the theory of karma relates to the sixth and the final element of non-violence. If one believes that all of life is interconnected and interdependent, then a logical expression of this belief is ahiṃsā. Although the Hindu narratives do not establish consistent philosophical theories, ethical principles, or conclusions, they demonstrate an implied regard and confidence in an individual's ability to reason ethically, structured through the principles of karma and ahiṃsā. Each situation places one in the center of a circle, with several options emanating in different directions. These options provide opportunities for action, opportunities for individuals to fulfill their dharma and influence the course of karma for themselves and others.

Modernity and globalization have influenced Indic traditions like other traditions. This globalization and modernization of Indic traditions is more prominent in the urban Indian organizations that are evolving like other NGOs to incorporate environmental projects in addition to various social and educational endeavors. They are not different from neo-Christian organizations in North America which are also "becoming green" in their agenda. Urban neo-Hindu organizations have juxtaposed diverse Indic traditions into a concrete religion with definite laws and rules to summarize the entire tradition. Following Sontheimer (1989), this urbanized version of Indic tradition can be called the "religion" of Hinduism since they strive to present a monolithic version of Hinduism. In addition, since this religion is not the daily routine "way of life"

---

[7] The diverse and "contextualized" ways of Indian behaviors are noted in one of the most widely cited essays by A.K. Ramanujan: "Is there an Indian way of thinking?" Marriott (1990). Ramanujan cites Hegel's observation as an example: "While we say, 'Bravery is a virtue' the Hindus say, on the contrary, 'Bravery is a virtue of the Kshatriyas.'" Also, see Narayanan (2001) in which she demonstrates the diversity of the Hindu traditions in which local custom can supersede the texts of dharma as done for centuries. She shows this using a case from the Indian Supreme Court, which prefers local customs and traditions over dharmic texts to decide whether a shudra can become an ascetic.

of rural Indians, they have taken environmentalism as a *separate* item in their agenda. Examples in this category are AIM for Seva, Art of Living, Brahmakumaris, Amma's Greenfriends, and others, as I showed above.

On the other hand, we have grassroots rural Indian groups and tribes such as Bishnois, Bhils, and Swadhyaya.[8] These groups continue to live the *dharmic* way of life in the sense that for them Indic traditions are part of their daily way of life and thus there is no such thing as "religion" in their lives as there is no separation of sacred from profane. Therefore, there is no such thing as environmentalism distinct and separate in their lives. Being dharmic brings them closer to practicing ethics to maintain the ecological order around them without being conscious of it. If Bishnois are saving animals and trees from invaders, they are simply living their traditions, not "protecting the environment" per se. If Bhils continue to practice their rituals in their sacred groves, it is their ancient tradition, not "saving the bio-diversity." If Swadhyayis are building Vṛkṣamandiras, they are simply expressing their devotion and reverence for all creation according to the teachings of the BhG, not "restoring the environment." Athavale once stated:

> We are trying to bridge the gulf between the haves and have-nots but we are not socialists. We are trying to make women aware of their strength, but we are not women-liberators. We are trying to improve the education system, but we are not educationists. We regard the downtrodden as the children of God and work for their welfare, but we are not social workers. We are mere *Bhaktas* (devotees).

To this list of Swadhyaya activities, one can add that, "Swadhyayis are planting millions of trees, working for water resources, and providing care for the cattle but they are not environmentalists, they are mere devotees expressing their devotion for the sacred nature."

From the above two categories, I have distinguished "religion" from "dharma." My point is that urbanized modernized Hinduism which is often included as a "World Religion" includes environmentalism with other social charity work. Thus, religion, ecology, and ethics emerge as separate categories in these organizations. On the other hand, the traditional, comparatively much less modernized Indian groups do not see religion, ecology, and ethics as separate entities. In line with the etymological definition of dharma, their *duty*, *virtue*, *cosmic ecological order*, and *spiritual* aspects of their lives are all intertwined just as dharma in its various definitions and meanings includes *duty*, *virtue*, *cosmic ecological order*, and *spiritual aspects of lives.*

Thus, my ecological perspective to research different Indian groups has helped differentiate two versions of Indic traditions as they are lived and practiced by

---

[8] Swadhyaya community has been compared with another "aboriginal" community, Pitjantjatjara at Amata on traditional tribal lands in the Central Australian desert. www.laetusinpraesens.org/docs/indiaoz.php (viewed on June 1, 2007).

millions of Indians in India and the diaspora. Scholars such as Balagangadhara have criticized the "religion" version of Indic traditions because, in their view, this is a reductionist and a poor imitation of Abrahmic religions.[9] On the other hand, urban Indians commonly criticize the traditional Indian practices of performing rituals and reject them as mere superstitions. Instead of falling in either camp, I simply note that diversity of Indic traditions allows both kinds in its vast and now globalized evolving forms. Neo-Hindu organizations such as the Ramakrishna Mission, Arya Samaj, and Brahmo Samaj had also evolved after the British encounter in the eighteenth and the nineteenth century in quite different ways from the traditional rural Indian way of ritualistic practices. They had criticized the traditional ritualistic way of Indian spirituality and instead adopted a more rationalized form of organized "Hinduism" largely as a reaction to Western modernity, as noticed by Halbfass (1988). In the same way, today's urbanized Indian organizations are merely responding to globalization in their own ways.

Ramachandra Guha presents three ecological utopias (2006): agrarianism, primitivism, and scientific industrialism. The first is the Indian model based on sustainable farming, the second is the biocentric vision of the "deep ecology" project, and the third is the "shallow ecology" project aiming to protect the ecology for consumption of the rich societies while ignoring the needs of the poor. Guha criticizes the latter two, prefers the first one for India, and eventually even seeks to synthesize all three, "This synthesis would take from *primitivism* the idea of *diversity;* from peasant culture the idea of *sustainability;* and from modern society in general, rather than scientific conservation in particular, the value of *equity*." What he means is that the "deep ecology" project's emphasis to preserve the biodiversity, Asian farmers' sustainable farming practices, and social equality championed by modernity can all enrich the current ecological discourses. He then goes on to advocate Social Ecology as a new discipline that combines all his three utopian visions and thus is the ideal ecological solution. Although I agree with his "Social" vision, my three case studies and several other studies by David Haberman, Ann Gold, and others actually bring in one important additional element. In their environmentalism, Swadhyayis, Bishnois, and Bhils are not really interested in or conscious about "primitive biodiversity" or "modern equity." They also are not environmentalists simply to sustain the ecology or even their livelihood. What really inspires the protective and reverential attitude toward ecology are their traditions rooted in their dharma. This important dimension continues to be ignored by "socialists" such as Guha and others.

In my observations, I found that Indians describe these different categories by their common vernacular term dharma or dharam. Just as Mahatma Gandhi in the nineteenth century and Athavale in the twentieth century made use

---

[9] http://www.india-forum.com/indian_culture/Why-Understand-the-Western-Culture-044.html (viewed on June 6, 2007).

of dharmic traditional symbols and myths to inspire their followers to take nationalistic and socio-spiritual causes, more Indians can be inspired to take the environmentalist cause if the message is rooted in dharma. On the one hand, it is true that like any other traditional societies, Indians are yet to wake up to the problems of ecological disasters. Most sects, castes, and other traditional Indian groups continue to practice their nature-worshipping rituals without being mindful of the ecological connections. Yet, on the other hand, despite being the second most populated country on the planet, India continues to boast of the richest flora and fauna on the planet, the biodiversity that has been preserved for thousands of years. My study of Swadhyayis, Bishnois, and Bhils provides us with three different kinds of models connecting dharma and ecology.

To be sure, all these three communities present considerable differences from each other. While Bhils do not have a charismatic founder with a set of teachings or guidelines, both Bishnois and Swadhyayis derive their inspiration from their gurus. Even between Bishnois and Swadhyayis, we can note that the Bishnoi community also started like the Swadhyaya community, as a "New Religious Movement," but over a period, like many other such Indian NRMs, Bishnoi itself has become a caste-group with strict marriage and other usual caste related customs. Swadhyaya being a relatively newer community has so far tried to maintain as a loosely formed community movement rather than a caste-group. In its early period, Bishnoi spread like any other NRM, with new members joining by taking an oath to follow Jambheśvara's teachings. The spread of Swadhyaya can also be accounted to new members joining although there was never a formal oath-taking ceremony.

As Ramachandra Guha and others have noted, environmentalism of Indian communities has usually been inspired by the survival needs of the human society, different from the agenda of "deep ecology." For Bishnois, the need to restore and revere the surrounding nature was to be able to survive in the hostile desert of Rajasthan. When Jambheśvara recognized this need, he intertwined it with the dharmic traditions rooted in various Indian cultural symbols and myths, and his followers wholeheartedly embraced it. This paved the way for their unique environmentalism of sacrificing several human lives to save trees and animals as is evident from several such examples in their history of hundreds of years. On the other hand, for Athavale, the question was not so much about human survival, but it was the survival of Indian traditions and Vedic culture. Instead of turning it into a neo-Hindu kind of religious jingoism or chauvinism, he strived to build and strengthen the relationships among humans, nature, and the Almighty. All his prayogs are based on his idea of recognizing the "Indwelling God" in everybody and everything. Thus, for him, the question of survival became the means for implicit environmentalism, although his message was more proactive than reactive. Instead of protecting the trees and animals from poachers, Swadhyayis went on to build new tree-temples and other natural reservoirs and resources. This can also be compared with Bhils who did not create any new sacred groves but continued to maintain their ancient groves based on their faith in the deity associated with a particular grove. Just as Bhils preserve their several sacred groves based on different deities, Swadhyayis have their several tree-temples, recharge wells and tanks, and devotional farms based on "Indwelling God" in trees and nature in general. The successful maintenance and sustenance of these Swadhyaya prayogs rests with the conviction and enthusiasm of villagers to maintain the dharma of their own as well as the dharmic teachings of the Vedic sages (as interpreted by Athavale). So much so that all the tree-temples are named after different Vedic sages to honor and revere those legendary sages. Arguably being one of the few communities building new sacred groves, Swadhyayis have thus distinguished themselves from Bishnois and Bhils.

However, Swadhyaya still being a relatively newer phenomenon than Bishnois, much of their activities are yet to transform in their "way of life" or ethos as is the case with the Bishnois. For Bishnois, the entire region of Western Rajasthan is their "tree-temple" where they are ready to protest against anybody harming the trees or animals. It is remarkable that the community of Bishnois arose from the people relying on non-vegetarian dietary practices. However, due to the charisma of their guru and his teachings, their ethos was transformed to become reverential and protectionist of their environment. This is similar to the transformation in the Jain community of Rajasthan whose ancestors were of Rajput castes with non-vegetarian dietary habits before they adopted Jain lifestyles based on non-violence and other strict dietary restrictions. Although Swadhyayis show a devotional attitude for their tree-temples in the rural areas, one can only speculate on their ethos for the future. Will the charisma of their founder continue to inspire them to maintain their ethos to become active environmentalists similar to the Bishnois? Bishnois had a set of 29 rules given by their founder that often are seen like commandments to be strictly followed. Jains also have their rules that are followed by the ascetics and the lay people. However, this is conspicuously absent for the Swadhyayis. Although the prayogs given by their guru are maintained in the letter and the spirit, there is no clear list of dos and don'ts. Thus, there is a possibility that Swadhyayi environmentalism might be restricted to their tree-temples instead of treating the wider region as a comprehensive tree-temple. For example, it is not clear if a Swadhyayi, who moves from a village in Gujarat to a town in New Jersey, connects his reverential attitude to make his lifestyle sustainable outside the Swadhyaya sphere of activities. Do the tree-temple devotees buy only "green technologies" in their homes or offices? Will they protest someone harming a tree or animal similar to Bishnois (even when their founder's teachings are now more than 500 years old)? These questions are beyond the scope of this book.[10]

[10] I came across Chicago's "green guru" Aufochs Johnston, Assistant to the Mayor for Green Initiatives, who successfully evolved his environmental sensitivities that he developed in his childhood days in India to develop several sustainable projects for the city of Chicago. See www.believechicago.org/sustainablecity/2005/05/guru-of-green.

Extending this distinction further, I would argue that Swadhyayis, Bishnois, and Bhils are also different from the Chipko movement. The activists of the Chipko movement were conscious about their survival needs derived from their surrounding forests, but for Bishnois, Swadhyayis, and Bhils, protection and maintenance of their natural resources are more dharmic than economic. Thus, I agree with Ramachandra Guha to categorize Chipko and other similar movements as examples of Social Ecology while my case studies should be categorized under Dharmic Ecology, as I noted earlier. When Bishnois save a blackbuck, their inspiration is based on the dharmic teachings of their guru. Swadhyayis build new tree-temples and Bhils protect their groves with similar inspirations. Following Milton Singer, I suggest that India's traditionalism works as a built-in adaptive mechanism for making changes. Essentially, it is a series of processes for incorporating innovations into the culture and validating them. Thus, the charisma of gurus such as Jambheśvara and Athavale could incorporate their innovations and validate them based on the traditional interpretations. These examples seem to match Milton Singer's conclusion, "Indian civilization is becoming more 'modern' without becoming less 'Indian'" (1972: 247).

This brings me back to the discussion about the "great" and "little" traditions. The human relationship with nature is a "little" sum of their daily practices and everyday experience in India and elsewhere. On the other hand, scholarly and political discussion about this human–nature connection tends to be largely a "high" armchair exercise. What I have tried to do is to bridge this gap by bringing in the "little" linguistic and cultural framework of traditional rural communities. Instead of juxtaposing "high" external terms and categories, I have tried to describe the religiosity and ecology of these communities in their own "little" categories. Although I agree with Frederick Smith that "linguistic borrowing and influence cannot safely be equated with cultural borrowing and influence" (2006: 149), borrowing the Indian vernacular words into the Western study of Indian traditions is imperative for the "thick description" of Indian ethnographies. In the colonial times, when *bhadralok* Indians borrowed the English word "religion" for their language, they also unintentionally borrowed the theological framework to reduce Indian culture into neo-Hindu organizations. Frederick Smith cites Appadurai to demonstrate how Swami Vivekananda "largely succumbed to the androcentric, linear, protestant, and activist strands in the dominant colonial culture" (2006: 587).

However, the vast majority of Indians in rural areas have remained largely uninfluenced by the urge to "modernize" their traditions and "way of life" into "religion." Therefore, to study and understand Indian traditional societies, it is helpful to drop the bhadralok framework of "religion" and instead use the "little" terms such as dharma, which has long expanded from its limited śāstraic idea to vernacular multivalent roles within Indian society. Unless the ecological

---

html. It remains to be seen if the Swadhyayis will emerge as similar examples outside their native villages.

awareness is translated into a dharmic message, it will remain a distant voice largely limited to armchair exercise in political or scholarly discussions. For the majority of Indians who speak, think, and read in vernacular framework, terms such as "global warming" or "biodiversity" have limited appeal and thus "high" rhetoric based on these terms will have limited appeal. When these are translated into dharmic framework, as was attempted by Athavale and Jambheśvara, it turned into notable successes. Fortunately, even the urban modern religious communities, in both India and the USA, are also on this path of reinterpreting their theologies into ecotheologies. The sooner these textual and intellectual "high" exercises can become widespread "little" practical movements, the better for our endangered planet. After all, what has sustained our planet for thousands of years is the sustainable need-based usage of natural resources,[11] not the greedy exploitation of the last few centuries in our drive for modern luxuries and comforts. Historically, population and consumption of natural resources of Asian civilizations have always been many times more than Western civilizations and yet it is the Westernized notion of progress in last few centuries that has endangered our planet (Guha 2006: 5). Unless we reform and expand our idea of progress by mixing ideas and concepts from the local communities, all ecological

[11] Even with a rapid growth in its population and a growing economy, India remains as one of the world's top 12 megadiversity countries with a rich variety of biological community types that includes coral reefs and alpine meadows, rainforests and desert scrub (McNeely et al. 1990). India also has the world's largest environmental movement with about 1,000 NGOs (Peritore 1993; Chapple 2000). According to the biennial State of Forest Report published by the Indian government in 2001, the forest cover increased by 2,000 square kilometers. The increase was in both categories, dense forest (40 percent canopy) and open forest (10 to 40 percent canopy). Also for the first time, the forest cover crossed 20 percent (up from 19.39 in 1999 and 19.27 in 1997). As per the 2003 report of the Forest Survey of India, India's total forest cover rose by 2,800 square kilometers between 2001 and 2003. There is a net increase of 21,000 square kilometers. Forest and trees cover 23.68 percent of the area, or over 778,000 square kilometers – a net increase of 21,000 square kilometers, 0.65 percent more in area (*Times of India*, July 21, 2005). This confirms the UN's World Forest Report in 1999 that reported that India was the only developing country where forest cover was increasing. However, according to the FSI report released on February 12, 2008, about 700 square kilometers of forest cover was lost between 2003 and 2005 and 20.6 percent of India's geographical area is under forest. Although some of this data has been disputed, as told to me by Paul Robbins in an email, there is evidence that the rate of decline in tree cover in India has slowed significantly since the mid-1980s (Saxena 1999). Rajasthan state has registered an increase of 3,478 square kilometers of forest cover during the assessment period between 1987 and 2001. In addition, the area of desert in Rajasthan has actually decreased from 1880 to 1980 (Haynes 1998). As a small example, initiatives such as Maiti Ritual have resulted in Uttaranchal now having 67 percent forest area in contrast to Uttar Pradesh, which has 17.5 percent. In 1950, Uttaranchal had 58 percent forest cover, which had dwindled to 36 percent in 1980. In this ritual, bridegrooms are required to plant trees during the wedding ritual. Around 500 villages practiced this ritual as of 2002. See www.comminit.com/en/node/121985.

rhetoric will remain "high" for the "little" communities around the world. In February 2006, Majora Carter, founder Executive Director of Sustainable South Bronx, an ecological NGO in New York City, harshly criticized Al Gore for his "top-down" environmentalism and argued instead for a "bottom-up" approach involving the enthusiasm and experience of the local communities.[12]

People can use their traditions to justify a position one way or the other. The key is to support and encourage those who try to make a positive difference in their interpretations and practices. Tradition is but one tool, but one that should not be ignored. For believers, such modeling can change behavior, even if it seems tiny in the face of huge forces going in other directions. Many other cases worldwide should be examined critically for the lessons they hold for ways to get people involved. Overall, I would like to end on a cautious optimistic tone for India's ecology, especially with the increasing combination of Indian traditional knowledge with the role of NGOs and governmental regulations (Sawhney 2004).

---

12 www.ted.com/index.php/talks/view/id/53?gclid=CKDUopybg5ECFQGnPAod P255HA (viewed on January 19, 2008 when NPR broadcasted an interview with her).

APPENDICES

Appendix A

Translation of Jambheśvara's Śabdas[1]

O purohit, you should identify that guru who has identified the ultimate guru (God). That guru preaches about the dharma. That who is worthy of becoming guru is of natural character, embodiment of brahma, self-consumed, and endowed with Vedic qualities. These are the identifying adornments of the guru. Consider only that guru to be manifested who is established by the six-fold darśanas and who establishes the worldly pot by his own hands. You experience and have darśana of that guru but the path to reach that guru is difficult. He is beyond speech and their voice becomes speechless. All rudras are included in him. That guru himself is quite content but nourishes others and the entire universe. The voice of that guru is soaked with spiritual wisdom. Some pots are unclean but when they are roasted in the fire, they are purified and then the cow's milk is collected in them. In the same way, with the holy company of the guru or by chanting of God, a lowly person can achieve greatness. However, diluted buttermilk can provide neither *ghee* nor store milk or pure water. That is, without the refuge of guru and God, one cannot benefit from worshipping other gods. Therefore, one should search for wise guru and pray for him. That guru destroys the illusion just as *shaan* destroys the corrosion from iron. Mind is purified with the holy water of preaching. Only *sadguru* can remove the suffering of the mind. This is the true identification of a sadguru. Without the yoga-līlā of Kṛṣṇa, raw clay could neither store water in the past nor never will in the future.

I have neither shadow *chhāyā*, nor illusion *māyā*, nor blood, nor flesh, nor any other bodily elements. I do not have parents. I am self-illuminated, i.e., I am self-produced. I do not weep, cry, or get angry or distressed. I neither have sorrow, nor am I cursed or curse anybody with any curse. I am present and yet unaffected in the three worlds *triloka*. I am unique. We only remember that by whose chanting cycle of life and death will be broken. Nobody knows my beginning. Worldly people merely have a hazy guess about me. Worldly people are suffering from three kinds of pains, physical, spiritual, and natural, one should cover these three pains. I am the creator of the beginning and the end, and then who is the creator of me! I am Yogi or consumer of worldly matters or a light eater. I am a wise meditator and sustainer of karma. I am caretaker of everybody or a support of all like a reflection in the water (as the Sun reflects in the water, I also reflect in the world). Accept the compassion-dharma. I am a *Bāla brahmachāri*.

---

[1] My translation is based on Jambheśvara's discourses published by Suryashankar Parik (2001).

My body neither was massaged with alasi-oil nor was anointed with perfumes. I do not eat food, drink water, or consume anything else; what is my sustainer then? The 68 pilgrimages are within my heart, outside pilgrimages are merely for worldly affairs. All the creatures, small or big, take birth and die in my breathing. Their life and death takes time equal to my breathing.

When there was neither wind, nor water, nor earth, nor space, nor moon, nor sun, nor sky, nor stars, nor cows, nor bullocks, nor the illusion created by *Māyā*. There was neither love, nor mother, nor father, nor siblings, nor any other relationship, nor gentle persons, nor any nepotism, nor family. There were neither the 8.4 million species, nor 18 kinds of vegetations, nor seven kinds of undergrounds, nor *sheshanāgas,* nor salty ocean. There was neither dead or alive species, nor man-woman couple. Neither was money, nor wealth, nor pride, nor horses of flying speed. There was neither shop, nor city, nor market, nor forts, nor palace-gates. There was neither any desires, nor habits. It was just one *Niranjan Shambhu*, without any *Māyā* or it was just dense darkness. O creature of the world! Which time are you asking? I am going to discuss about the 36 eras and even the next 36 eras beyond them, which does not have a final ending boundary. I existed then; I exist now and will exist in the future. Tell me if I should discuss that era.

O guest to the world! This is my supernatural voice that does not remember born creatures. Nine incarnations and Nārāyana have stabilized in me alone. Why should one chant the chanting sage, ascetic sage, or other renunciates or sages? They are all creatures who took birth in bodily forms. Why should one chant the eagle or other birds flying in the sky, or the creatures walking on the earth, or the manifested and hidden *kshetrapāls*? They are also merely less wise creatures. Why should one chant Vāsuki snake and *sheshanāga* also? Because they are also creatures that are born. Why should one chant the 64 *yoginis* and 52 heroes? While they are also born creatures. I chant only the independent Shambhu who has neither mother nor father. His body has neither blood, nor other elements, nor heat, nor cold. He is creator of all and devoid of his own death but the origin of all the creatures. O guest to the world! This is perennial voice; I do not chant the born creatures.

I am pervaded in all the stations as an eternal spark. I have selected jewels and pearls. I am an explorer of the truth but you are not aware, that which I explore is *Allāh*, pure, unwritten, unborn, original, self-absorbed, indestructible, and what not! How to describe Him? I have obtained the knowledge about Him not by sitting with someone but have obtained Him by devotion and continuous chanting. I am a Hindu by birth, a yogi by tolerance, a Brahmin by action, darvesha by heart and mullāh by indifference. My intellect remains always like this.

Being a Hindu, why did you not chant Hari? Why did you wander your heart in ten directions? Why did you cut vegetation on Somavati Amāvasyā and Sundays? O fool! At the time of eclipse, why did you drive a cart, on nirjalā Ekādaśī, why did you sleep on the bed? As long as there are no homa, chanting, ascetic practices, etc. not performed in your home, know that the fertile cow is away from your home. If you have done false work, it will lead to destruction of your merits. O deluded creature! You merely indulged in gossiping. You did not chant the holy name of Viṣṇu. Everybody likes the self-appreciating talk, who gets convinced of the truthful sharp speech? At the appropriate time of awakening of the heart, you did not awake from your heart but kept apprehending being doubtful. Even in cold times, you woke up only to cool something and kept running around for selfish motives during the day. Even at the dusk, you did not remember Viṣṇu; did you leave any mark by your behavior? Did you earn anything special? In your lethargic delusion, you did not identify God. If you did not obtain the knowledge of *Pāra-Brahman*, the *Nāgā* sage also could not achieve the *Yoga*. Who did not die for obtaining Paraśurāma, certainly wasted this human body.

O qāzi listen! O mullāh, listen! O goat-killer butcher listen! On whose authority, do you kill goat and on whose instruction do you kill sheep and cow? You suffer intolerable pain if a thorn pinches you, will the live animal not suffer the same pain? You knife-wielding Turk claim to enter the heaven even after eating the inedible! Only milk obtained from grass-eating beast is suitable for consumption. Why do you ply knife on such a useful animal? You remained blank even after getting education!

For one whose heart is pure, Hajj of qābā is close. Then why do you shout to obtain Him? Bullock is dearer than one's brother, then why do you ply knife on its throat? Without proper identification, Khudā will always remain separate, what kind of Muslim are you? Kāfir has failed from his promise, has spoiled his path, and adamantly wants to see God. If the God could be identified in the way you shout toward the west, heavenly aircrafts would arrive at the time of your death, why do you shout on the walls of the mosques? Why do you destroy cows? If cows were to be killed, why did *Karim* graze the cows? Why do you eat milk and curd and why do you consume *ghee* and buttermilk? Having consumed all the dairy-products, why do you now consume meat and bones of cows? Why do you kill them and drink their blood? O kāzi listen! O mullāh, listen! Who is dead in this? Who tortures the animals, will suffer immensely at the time of death.

One who sacrifices oneself from beginning with *Rahmān Rahim* in his heart, *Rahmān Rahim* will obtain him with His mercy. Only benevolent actions should be done by the body, only duties and noble actions should be done by heart. Duty should be kalamā and speech should be the Qurān, this is one's *kula*, family status. If you seek God within your own heart, you will become like darvesha and will become true Muslim. Look! Peer, elders, and the community of Muslims read this in scripture and this is what it instructs, God exists in everybody's heart, mine and yours too. One who thinks in this way obtains the mercy of God. O *Miya*! If you walk on this path of spiritual practice, you will reach the heaven.

If your heart is pure, hajj and qābā are near you, then why shout loudly for Him? Practice true devotion with pure heart and perform your duty truthfully just like reading *namāz*. Even if you earn little in this rightful earning, be satisfied with it. Hey! Why do you kill innocent animals? God will ask for the account of your actions! You know how to ply knife on them but do not know their pain. You just read *namāz* blindly. A beast that grazes upon the forest grass and provides milk to you, only its milk is pure and to be consumed, why do you ply knife on its throat? Even after reading the Qurān, you are empty of qualities, why do you shout from the walls of the mosques and markets? Your motive is false. Your action is evil and your *namāz* will remain useless. How do you purify yourself? How do you remove your sins? How do you concentrate on God? How do you identify God? O Mullāh! The mosque is in your mind, read your *namāz* inside, does not He listen that you shout loudly for Him? You have not known God but have only identified the world. So what if you have had *sunnat* (circumcision)? O ignorant! If you do not identify your moral duties, you will certainly be put in the hell.

O qāzi! Do not quote Mohammad to support your animal-violence. His thoughts were quite complex and different from yours. His was the sword of knowledge that removed the sins of his people, not the iron sword! He liberated 100,000 sages with him. He was grateful and faithful to God but you are dead.

O fool! Why did you waste your human life? Why did you burden the earth with your weight? In your youth, you neither performed homa, nor chanting, nor austerities, nor you identified the guru, nor you obtained the true path, thus you wasted your human life. In your youth, you did not awake for God. Neither you woke up in the cold morning, nor you remembered Viṣṇu in the evening, and you suffered heavy loss. You did not earn the real wealth, your body is perished, and your life is not stable. Everyday your age is reducing. Without remembering Hari, your breathing is also being wasted. One who did not chant Viṣṇu-mantra, they are stigmatized, lose their family status, will be born as lowly creatures, they will suffer heavy burden on their shoulder, they eat extra and their blood and flesh are wasted. Who does not chant Viṣṇu, will be born as sheep in the villages, pigs in the cities, and weight-pulling donkeys. They will attain the body of that bird that remains silent in the night but has to open his beak in the garbage in the morning to search for food. They lift the heavy burden of sorrow; they will not be able to cross the ocean of *bhavsāgar* and will enter the dark hell. Neither herbs, medicines, nor any magical formula will work there. Time spent in this way is as futile as something fallen from a mountain. Do not blame Viṣṇu for your plight; it is the result of your own sins.

My preaching is equivalent of Veda but who knows this philosophy? This cannot be written in books and scriptures. Search for the highest spiritual element in my words. Why does lion kill deer and tiger kill cow? There is something special about Kṛṣṇa-Lilā. Bitch and its puppy, being cowards cannot become tiger. Without the grace of Kṛṣṇa, eagle's nature can never become gentle and noble. Donkey's voice cannot become sweet and without Kṛṣṇa's grace, dog cannot become serious. Bald woman does not give birth to bald child, without Kṛṣṇa's grace, beer can never become pure. Cat's tongue cannot be contented. Without Kṛṣṇa's grace, dry heart cannot become emotional, chicken cannot become peacock, and thorny clothes cannot become silky ones. Coconuts cannot grow on Neem, diamonds cannot be found in a mine, *Indrāyana* creeper cannot become *Nāgar*, *Babuli* tree cannot become khejari. Beast cannot fly in the sky, Bhil hunter cannot have compassion, pig-cub cannot become elephant, and dwarf cannot become tall. Even if crow has same color as cuckoo, it cannot become cuckoo and without Kṛṣṇa's grace, a crane cannot give birth to a swan. A wise person is delighted listening to my preaching but an ignorant gets irritated.

Accept my subtle words. I have not uttered these words to characterless people. O creature! Water that plant whose fruit and root is always sweet. However, you are merely lost in the leaves without any quest for the root. Why do you water the wrong plant? Chant "Viṣṇu-Viṣṇu" so that you can eliminate anger, desire and other such vices, this is the real purpose of life. O creature! Search for such profound spiritual wisdom that is beyond attributes. Guru is as great as a mountain and as pacifying as water, i.e. He pacifies his disciples with His knowledge. Among dry-fruits, he is like extremely sweet fruit. His heart is kind and satisfying. He is the knower of the astrologer who knows past, present, and future. He fulfills the hopes of the hopefuls. Serve that guru who can cross your lifeboat as a boatman.

I have transformed iron-like people into gold and established them like gold-jewelry. In this way, jāts are purified with the grace of Kṛṣṇa. By your good fortune, you have found me and like a boatman, I will cross your boat over the *bhavsāgar*. How can iron float on the water? Similarly, you can also float only with the help of the right company. Even without right action, if one will sit with me, one will cross over just like iron floats on the water with the company of wood. Worldly people regard the naked and arrogant animal-sacrificers as renunciates and get deluded.

Right and deserving speech is my wife, thus my mind has become wise. If we observe with discrimination, there is no difference between the breathing, flesh, blood, and soul of man and woman. Should we accept the statement of a deluded person whose knowledge is doubtful? O guest to the world! I do not desire anything from man or woman. In fact, *Brahman* is omnipresent. One who searches for profound guru is *subhyāgata*, the real seeker, but one who wanders to different houses is a beggar. Beggar can seek alms if he has accomplished the highest spiritual wisdom. One who has evils such as attachment, hatred, doubt, etc., who can call him *sālhiyā*, the pure seeker.

Who says with false-pride that he has known the God, has not known anything. One who humbly says that he has not known anything has indeed known something. "I do not understand the God's indescribable tale," one who says this, his voice is divine. Those who are wise consider their speech as the highest stage of knowledge and those who are educated consider their knowledge in mere story telling. Just as peacock cries looking at his ugly legs opposite his peahen, the so-called educated people also waste their lives trying to show off their knowledge. Yogis are anxious to practice such austerities that will make them attain higher stages, and fools keep crying and keep trying to accomplish worldly things. Understand the gist of death; it destroys all in the field. Several incarnate divine beings cry for the people who do not understand this path.

If worldly people are devoid of herbs and medicines then why do doctors die? O creature! Search for such God whom fools cannot search. The unattainable God will be attained only by humble people. I reiterate that only they will have something.

I exist in the pinda, in the universe and in the heart of every creature in manifest and hidden forms. I am called eternal in the infinite eras. I have neither father nor mother. I have neither *māyā*, illusion nor *chhāyā*, shadow. I have neither a form, nor any lines in my divine form. I am unapproachable both inside and outside, I am unbounded, attainable by only those who can experience God. The 68 pilgrimages are inside the heart, only rarely can somebody visit the internal pilgrimages.

Wherever there is lack of compassion, there evil actions will be done. Wherever there is no respect, how can there be heavenly bliss? Those who lack the light of knowledge will never be able to attain the mokṣa. Wherever God is not searched, there the obvious is gross. Wherever the secret of God is not known, on what basis will the heaven be attained? Wherever there is a false pride, one will obtain neither heat, nor cold. That is, such people will be devoid of both calling and peace, and they have wasted their breathings.

For that yogi, for whom two opposite poles of dawn and dusk are the same, where is that pure water elsewhere? Take any instrument; there are only two deep instruments. One instrument makes sound when rain pours, the other makes sound while making buttermilk. However, for that yogi, opposite poles are the same, in that yogi's sky and underground, the vibration of *soham* (I am That *Brahman*) echoes. He found the real treasure in his *samādhi* but concealed it from the worldly people busy in insignificant debates. Even great demons were destroyed in debates, just as the bees are entangled in the closed flowers in the night. O creature! You may or may not know but Yama is an enemy of your life. The sound of Bher instrument is audible within one or two yojans. Thunderstorm is audible within five or 10 yojans. O creature! You consider only guru's speech as the truth. If you speak that speech which can be heard even to the remote place, you will be benefited. However, that is for people of noble qualities, and that is unbounded.

O kings and emperors! Pleasant breeze blows and water falls from the sky. However, if someone worked on barren land, he will not be able to grow anything. Similarly, one who worked in productive field gets good benefit. Someone produced grapes and others produced sugarcanes. Somewhere Neem or elsewhere Dhāk also grew. Elsewhere *Indrāyana* creeper or other wild plants grew. What you sow, that you reap. If the root is rotten, branches are rotten, leaves are rotten, fruits, and seed are rotten, then why blame the water? If people are devoid of auspicious actions, their fear will not go away. Some people have come to this world like a bird or a bat or an owl in whose heart there is neither knowledge, nor light, nor liberation, nor freedom. Their actions are like this, then why blame water?

Who is guided by guru, overcomes the fear of death but those who are ignorant of guru's teachings are scared of death. Those who found the true guru, were shown the true path, were purified of all delusions and even death is benevolent to them. They get bright new bodies glittering like jewels and eventually get liberated and attain mokṣa. Mokṣa is attained by loving and chanting Viṣṇu. Concentrate your mind on Viṣṇu, chant Him, and do not slander anyone. Bow your head for Viṣṇu, and accept His refuge. If you follow my statements, then this is the teaching of guru, you will attain mokṣa. Living and dwelling are fine stations but beyond them is mokṣa-position. Through which thought does one attain that? If adorable is bowed, pardonable is pardoned, digestible is digested, doable is done, and thus trained one can reach the mokṣa-home. Mokṣa "*ratan-kāyā*" (jewel body) is a replica of truth. One can attain this by the grace of guru, by highest spiritual wisdom, by practicing dharma, by fine character, by abstinence, and by pleasing the true guru.

At the high pedestal of being a sage also, one can practice fraud and deception. Only rarely one knows the simple and pure path to know God. One who walked on the simple and pure path attained the divine abode having left the mortal body.

O emperor! In your pride of power, do not get attached to the world that will be dispersed just as wind disperses dark foggy clouds in the sky. Dense clouds get scattered several times in the sky and this world is not loveable. O deluded creature! You did not remember Viṣṇu, why are you forgetting your death? In front of my eyes, devas, demons, etc. have perished then how can something be permanent in Jambudwipa? That will be destroyed. Foggy rain cannot bring water in the rivers and ponds. Swan of your soul will fly away and your hopes of life will be wasted, then your body will become widow and then even this widow will perish and turn into ashes that will be blown away from wind.

Filling your stomach to the brink does not bring any special quality; it is like filling garbage in the storage. This extra food forces the eater to carry additional burden of their flesh in their front and back body parts and they forget their real purpose of human life. Just by age, one can neither become great, nor attain mokṣa. Just by getting birth in high-class family, one does not become superior; superiority comes only by practicing fine qualities. Just by looking at gorakh, one cannot become siddha; only by following that path can one attain mokṣa. O seekers of goodness! In this Kaliyuga (dark age), be alert. You found the true guru who showed you the true path and removed your illusions just as the sun removes the darkness of night after dawn.

Reading the Vedas and Śāstras on paper or listening to them is nothing. Some other people merely bow to wooden and stone idols. Paper books are hollow and the songs sung by them are nothing. From which side does it go inside and comes outside, neither mother knows this, nor father. If the soul enters through the nose etc., then how was the body formed inside the ovum and how did the consciousness arise inside that body in the fetus? Listen! O qāzi! O mullāh in mosque! O hermit! O pilgrim! Through which way did the jiva enter the fetus? It happened just as sound made from the bronze pot merges with the bronze pot. That sound came from neither outside, nor it went outside. The same process occurs in conception. Jiva comes in a second and goes away too. This is as natural as change of season from winter to summer to rainy season. Just as Vibhīṣaṇa transferred the secret knowledge about Lankā to Rāma. In the same way, you also understand the secret of life and that will be your victory. As the sesame-seed loses its value, after oil is extracted out of it. By wisdom, by meditation, by reverberating the chanting, by studying the Vedas, one can achieve spiritual wisdom. Karṇa, Dadheechi, King Shivi, and King Bali also obtained the fruits of their karma. King Harishchandra, his wife Tārā, and their son Rohitāshva sacrificed their bodies. Without chanting Viṣṇu, your life is wasted, just as desert plants Āka and Kheep are wasted drying up in the forest. Similarly Qāfir, atheist, is wasted. White cloth and cotton can accept many colors but black cotton and evil person cannot be colored with any color. Just as Lākh does not have any value before being colored by majeetha, a person has no value without being colored with spiritual wisdom. Sometimes even the evil person leaves his home in search of guru but finds a fraud and gets misguided. Naturally, sweet fruit can be obtained by meditation but fraud gurus and evil disciples cannot have that. Full of impurities, fraud and desires for women, such a person can never reform without the teachings and grace of true guru and without reform can never attain bliss and peace. Such a person is blind even when seeing, is deaf even when hearing, his living is merely completing years, he is far from real life.

Fish and crocodile move under water but their navigation pathway is not seen to us. Similarly, to know the supreme wisdom, one needs to enter the spiritual realm. It has neither beginning, nor end. It is neither near, nor far. It cannot be known since it is infinite and unbounded. What can I say about its profundity? It can only be experienced. It has neither a form, nor any line, nor any tradition, nor any footprint, nor the death. It can also not be found by the 52 heroes. Only fish knows its way in the water where it dwells. Similarly, only a siddha can know the way of God, worldly people can merely talk about it.

Illol Sāgar

Infinite number of people have benefited from the teachings of guru. Salty Ocean is limitless. Infinite creatures have taken refuge in these teachings and they will cross over the ocean using the teachings. Guru inspired them to establish *Kalaśa* in the evening, in the night and in the morning and thus they will cross over. The same guru has arrived in the *marubhumi*, desert of Rajasthan, wearing an ochre cap. Become one with me and follow my teachings. One who unites with God will certainly remember Hari Viṣṇu in his heart and will attain the bliss abode and will be welcome by devas. There were musical instruments playing at the abode of Hari and by the grace of Kṛṣṇa, I incarnated in Jambudwipa in which farmers reside in all four directions. I came and warned Sikandar Lodi. He accepted the truthful character and started living a truthful life. I provided refuge to orphans, guided the misguided ones, and helped reach the destination those who were on right path already. Earth is my meditation, vegetation is my abode, my light is reflected in the space, Sumeru Mountain is my pillow, other mountains are my leg-rest, and my desire is my quilt. These four eras of this universe have occurred nine times, this repetition of 36 has been occurring in dark. However, I experience light at dawn, the beginning of universal cycles. Following the ideal of 33 koti devas, I have come to rescue 12 koti creatures chosen by me. Having incarnated here, I have established love in the hearts of the living beings. I am the emperor of high mandalas. I had stirred the ocean using Meru Mountain with Vāsuki leading it. In the Paraśurāma incarnation, I killed all the kṣatriyas. When Sītā was returned, I was present there. I had pierced the 10 heads of Rāvaṇa. I am discoverer of the true seekers but you do not know that I play dice as well. I defeated demon Kansa in duel. Karṇa entered Kunti's fetus but this was kept secret by me. Just like the *majīthā* hides the earlier color and makes it red. Similarly, Ker, a green desert plant, turns red after ripening but its real taste is known after tasting it. Similarly, mere outside makeup cannot make someone a real sage. You have experienced neither yoga, nor pleasures, nor chanted Viṣṇu. Why do you churn the chaff without the grain? This will not make you accomplish anything. I am a detached yogi and cannot be deluded by worldly desires and illusions. I accept anybody coming to me in any form. I like those who are truthful. My mind is mudra, my body is *kanthā*, and I have completed the path of yoga. I gurgled with the water of seven oceans but have neither drunk it nor remained thirsty. I have neither desire, nor hunger, nor thirst.

Kunchi Wāllāh

Yama doot has arrived, has bounded the jiva-ātmā to take it, and is asking it, "what karma it has performed?" Jiva-ātmā started shivering severely and got agitated. Neither its father nor mother can help it there. Only good karma comes along it. O Swāmi! I used to bow down to wind, water, sun, and moon. I used to follow the path of Viṣṇu and devas. O Swāmi! I used to chant your name day and night in this unreal world. God verified its good and evil karmas and joined it with devas. One who desires such heavenly abode should adopt their path and chant Viṣṇu. This body is land and jiva-ātmā is farmer. If both are pure and fertile, wherever the knowledge will shower, it will yield good results for its efforts. Crop is collected from the farming yield and then the wind blows away the husk and only pure grain is left. Similarly, you can collect knowledge from everywhere and then the teaching of the guru blows away the useless information, only the pure knowledge remains and that can be used to transform your life and experience the bliss. How can this be nice, one does not fear from God but frightens others, does not perform good actions but asks others to do it, does not speak pleasant words but asks others to speak evil words, does not want to die but runs to kill others. One should practice before preaching. Whatever is to be done should be done before dying; one should not suddenly die before performing duties. For purity sake, why do not you take bath? For welfare of jiva-ātmā, one should take bath. Ones who do not do that, they will wander like ghosts. Impure jiva will suffer in the abode of Yama. Human body is like a jewel but one kept uttering evil words like a pig and will become pig. Who else can donate 50 kilograms of gold except Karṇa? Gālav asked for one cow, but except Karṇa who could have given fertile cow. Who else donated gold except Karṇa? Even today, who else is celebrated as the King Karṇa? Except Karṇa who else donated gold-teeth to Brahmin even after being struck by the enemies in the battlefield? Which resolution enables one to enter the assembly of devas? One who humbly bows to the appropriate place and is forgiving follows the complete dharma, only such person can enter the assembly of devas and can stay there permanently. That is the final farewell to return one's home. Through which qualities Vidur crossed over but Karṇa had to return? Vidur donated with sincere humility for guru so he was liberated from the cycle of life and death. However, Karṇa donated with egoistic pride that forced him to come back to human life. Wedding songs are devoid of meaning and only sound pleasant to ears; spiritual wisdom is not like those songs. Some people do not realize its profundity and waste their lives in their delusion. Those who are deluded are stuck into useless discussions, do not have proper character or thoughts, only know how to enjoy taste, are attached to fame, and when they die they cannot attain mokṣa.

O creature! Properly devote yourself to Viṣṇu leaving aside all the other demigods and ghosts. Just as watering the roots of a tree supplies water to all the branches, leaves, and flowers automatically, similarly by chanting Viṣṇu, origin of everything, pleases all the other gods and goddesses automatically. Do not worship evil ghosts and false gods; they will only make you suffer. By watering the root of all, one will get auspicious intellect and using that discrimination, one will perform good deeds in life. In this way, fear of world, fear of death will be removed and cycle of life and death will be broken. Just as tree grows from water, your good karma will grow progressively. Removing Hari, you did not follow the Hari authority, indulged in useless discussions, did not know service to devas, and cannot be saved from death from Yama. Being deluded, creature did not chant Viṣṇu, did not search for the root, looked only at the branches, and kept devoted only to that. Do not accept jewels without the sea-eater and pearls without oysters. Do not look for them in small holes. Death is bent on destroying the human by every moment, every hour, every day, every night, every fortnight, and every month. Therefore, it is futile to stick to sweetness of the worldly attachments. The attached person will be caught by them just as a fish is caught by fisherfolk's net. When the jiva will be separated from the body, the head will be broken forever.

If someone donates 10 million cows on a pilgrimage, half a million horses, yellow ritual-clothes made of grain, gold and silk, powerful elephants, follows strict discipline like Karṇa, Dadheechi, Shivi, Bali, and Rāma, and yet is unnecessarily argumentative, too arrogant, greedy for sensual pleasures, one cannot cross-over *bhavsāgar* without Kṛṣṇa's grace.

Who is not born in this world and who will not get birth in the future? Who has not borne the burden of worldly suffering? Who did not go from this world? Who will not leave this world? Who remained permanent in this world? Several have gone from this world in the past eras, what to speak of a tiny human in this Kaliyuga. God in the heart of baby in the fetus is no longer the same God after his birth. Human forgets the God inside his heart and corrupts oneself in the world. Whose is mother? Whose is brother? Whose is family? Worldly people die repeatedly but do not recognize God. O creature! Just chant "Viṣṇu-Viṣṇu," be alert, and do not get deluded in the worldly faults. The results will be depending on your destiny. The BhG is not a poet's poem but it distills the fine essence from the rest. O creature! You are deluded uselessly; can you know God in this way? When God is known, you get liberated from the cycle of life and death and mokṣa is attained.

By the supernatural power of Kṛṣṇa, it rains and fills ponds and rivers. Some swim across them but those who do not know swimming sink into them. Even others who are scared of water remain thirsty near the water reservoirs. Even when food, wealth, milk, yogurt, ghee, and nuts are available and someone remains hungry, then what is the use of such a life? With the goal of liberation, if someone dies in the battlefield as an instrument of Kṛṣṇa, then that is appropriate. If the tongue gets tired chanting of Viṣṇu, then one can do without such tongue. Chanting Hari-Hari makes you dynamic, then do not regret for it. O beggar! If you have obtained supreme knowledge, then accept the alms. Begging is not objectionable. Those who have obstacles of doubt and dryness, how can they be called sage?

Vyāsas, the preachers of the Vedas, preach the Vedas but do not have faith in them. They are attached to the worldly desires. Those who have not recognized the guru and are not devoted to the origin of the world, they are devoid of dharma and utter uselessly.

Although qāzi follows the Qurān's statements, he has not understood its message and is a qāfir, atheist. Who does not identify the real guru, utters nonsense.

Blacksmith makes pots from iron, brazier makes pots from brass, and potter makes pots from clay. Just as these three kinds of pots are not separate from their building material, the universe is not separate from Brahman. However, one who has not identified the guru and not focused on the origin, such a fool can propound any nonsense.

O weak-body human! Why do you kill innocent cows? Be forewarned of the sin from killing them. You show your arrogance toward God in this way. This is the wrong path. Who has not identified guru, has not focused on the origin. Uttering of such persons is gross.

Company of good people is good company, only shining color is good color, that is, company of good people brings shining color in a person. With such company, one can cross over *bhavsāgar*. Good process is that which makes one's life better. Good battle is that which makes one's life better and takes one closer to mokṣa.

In the seven undergrounds, three worlds, and 14 abodes, like the space, inside and outside, universally present, wherever I see, God is present. I found true guru, he told me the true path, and removed all illusions; nothing else is left to ask now.

O emperor, listen! O of yogis listen! O sheikh, listen! O of sufis listen! O so-called sādhu and siddha listen! This body is perishable, it is born, and it perishes. Those who do not know the creation and destruction of the body are ignorant.

O yogi! For which siddhi do you apply ashes on your body to appear like the sages? Why do you sit in the graveyard and serve the ghosts etc.? Just as pot kept upside down has never stored any water in the past nor it can in the future without the grace of Kṛṣṇa. Yogi, jangam, nātha, digambar, sanyāsi, Brahmin, brahmachāri, pundit, qāzi, mullāh, all these play their own tricks. However, whoever played trickeries without the teaching of guru will get opposite results.

King becomes extremely sad having lost his kingdom. Explorer becomes extremely sad if his exploration is wasted. If a householder loses his wife, he has to face many problems. Worldly people have to cry even if they lose their wealth here. At the time of reaping the crop, if a farmer loses it to birds or thunderstorms, then he is also sad. A yogi ascetic performing austerities becomes sad, if he falls for his sexual desires. Similarly, o fools! You should also cry looking at your wasteful life. You are not siddha and yet try to veil your weakness. There is no truthfulness in your life because it is not cheerful as that of a yogi. Your worries have made you weak that you are trying to hide with ashes. Yogi, jangam, chanter, ascetic, peer, do not misuse the meter used to weigh stones to weigh diamonds, i.e., the means which is not for knowledge or mokṣa, do not use it ignorantly. A yogi will remain yogi for all the eras and is a yogi presently as well. O yogi! You pierce your ears and torture your body but this is mere fraud, not yoga. You grow your hair and practice animal violence; this is fraud, not yoga at all.

Your rough bag and kanthā are rough, why do you suffer from their burden? Why did you renounce your home and family when you are not acquainted with yoga? You started stitching this rough bag and clothes and started growing long hair and beard. Merely indulging in useless discussions cannot help you cross *bhavsāgar*. You chant the heroes and ghosts. Why do not you search for the *tattva*? O yogi! You have a stick in your hand but you have not restrained your mind. Your head is shaved but your mind is not purified. If your mind is purified, you will not indulge in worldly desires. Because of your delusions, you indulge in useless discussions and your faith has disappeared. You are not practicing compassion for creatures.

You cannot even stitch your clothes with equivocal heart and mind. Without focusing the mind and heart, you cannot travel on your path. With dilemma in your mind and heart, you can neither narrate a story nor listen to it. It is extremely difficult to obtain the goal for such a person. He can become neither guru nor disciple. He cannot regulate his time and cannot obtain God. Without focusing the mind, one cannot put thread in the needle-hole. With such a mind, your fate does not favor you and you cannot discover any secrets. Doubtful mind cannot follow discipline. An unstable minded person cannot perform even the worldly activities. It is impossible for an unsteady mind to reach heaven. O rāval of jogis! You wandered here and there. Without knowing the reality of yoga, what did you obtain? Why do you travel to foreign countries? Why do not you obtain the real education? You do not deserve to practice yoga. You neither enjoyed worldly pleasures nor followed the path of guru. Why do you churn the husk without the grains? O avadhu! Just as without proper footwear, feet are pierced by thorns, without proper clothes your naked body has to suffer. If you did not experience Sachidānanda Brahman, mere renunciation of clothes is useless.

Serve only that yogi for whom mind is mudra, body is ochre robe, who has stabilized his bodily fires. By pleasing him, you will cross over *bhavsāgar*. Those who are called *Nāthas* have not conquered death. They keep dying and taking birth in different species. I am *Rāval*, I am yogi, and I am emperor. Whoever comes to us with whichever intention, I establish him properly. Only truthful people are truthfully established. I neither hide their sins *pāpa*, nor grant merits, *punya* to them. Do not delay in performing worldly duties with proper living. Only while living, you can control your karmas; afterwards you will be rewarded based on your karmas. I am between heaven, *svarga* and hell, *naraka*. Those who come and meet me, I reform them and take them to higher abode.

Body is my bag, yoga-established mind is my ochre robe, breathing is my musical instrument. Restrain the mind-deer, apply it for yogic practice, and disregard the external showoff. I am yogi, I am truthful and restrainer of mind. O devotee of Adinātha! Control your five prānas and nine bodily apertures and practice yoga.

Do not chant "Lakṣmaṇa-Lakṣmaṇa" and confuse people. That Lakṣmaṇa was one, who fought fiercely, killed Rāvaṇa and conquered Lankā. He was emperor of three worlds; your Lakṣmaṇa is a common person, not a king even of a village. That Lakṣmaṇa controlled 8.4 million species; your Lakṣmaṇa cannot control even one. That Lakṣmaṇa was full of fine qualities; yours is indulged in useless discussions and delusions. Your Lakṣmaṇa does not have any quality then why should one bow to him?

O Avadhūta! Burn your mental defects, develop fine qualities, keep them always alive, and practice abstinence. By practicing strict celibacy, you can have Darśana of guru.

Others breathe just like you and have flesh like yours then why do you discriminate people? By good fortune, whose mental defects have been removed, he does not discriminate. The God that appeared in qābā is the same that appeared in this desert. Those who are obstacles in chanting Viṣṇu are atheist and evil people. Being Hindu, those who just take holy-dip in pilgrimages and perform rituals for their deceased ancestors will remain empty. Those yogis who shave their heads, wear earrings, and perform rituals will also remain empty.

Turk who just goes for hajj and merely reveres qābā is also lost. Many may be awakened in this world, I am such awakened in this desert. Whose knowledge, vibrations, character, speech is beyond comprehension, is Lakṣmaṇa.

Who keeps seven undergrounds, whole earth, and its world in his heart but does not avoid punishment to thief. Allāha, Alekha, Adāla, Ayoni Shambhu has adopted body of wind. Maya dwells in body, maya has dualism, dualism leads to ignorance, and God resides inside such ignorant jiva. Yogis do not get hungry, they can survive on air and do not need grains.

Throw out your attachments and sustain the difficulties. Such a person is a real king and his position is incomprehensible and infinite. Only the courageous can stay in the battlefield. Having the sword of knowledge in the hand, who can be my enemy?

Guru distributes diamonds whether you take them or not. However, do not blame guru if you are unable to receive them from guru. Wind, water, earth, cloud, vegetations of 18 species, established mountains, sunlight, and even abodes beyond these, all take refuge in guru. Ocean can accept fully filled rivers, streams, and ponds, similarly guru can accept the whole universe. The 99 different kinds of sensualist kings renounced their sensual pleasure and became yogis by the preaching of guru. They visualized guru, practiced yoga, and renounced their bodily attachment. Look! Work but do not strain much, jiva will leave you. In the earlier era, deer reached heaven by doing good karmas. If someone dies in living form itself, then such person will attain devas. Some people are of hybrid varṇa, evil jiva, unpleasant speech, and are arrogant and proud; they will die with extreme burden of their sins. O human! By sleeping in your ignorance, you wasted time with free hands. Like an inert stone, you wasted your life. You are lost in your delusions and wasted your life. What is that desire that makes you lose your life? Even after guru forbade you, like a blind person you tied yourself in the cycle of life and death. Without the grace of Shyāmasundar, nobody can cross over *bhavsāgar* even sitting on Pāras. O jiva! Accomplish your life with good karmas.

Before dawn, at the dawn, after the dawn looking at the Sun, in the holy time of Viṣṇu, and at the time for chanting Viṣṇu and Kṛṣṇa, why do you utter nonsense? When the eggshell of your body will be broken, time will be lost and good opportunity will turn into missed one. See the supreme person beyond me from divine eyes. Both of us have same light, my mind and breathing is by Him. Guru practices abstinence and we should learn from Him. He easily breaks the cycle of life and death.

Listen! You were walking on the footprints of rabbit and you found an elephant. You were searching for fox and you found cow. Listen! You used to dig the roots of a wild bush and suddenly you found a fruitful tree. You used to work with clay and now you found gold. Similarly, by your good fortune you found a true guru. A walking elephant and its rider do not get affected by barking dogs. Similarly, a refugee of guru cannot be affected by anybody else.

If unworthy person is given a donation, it is like stolen by a thief in a dark night. Thief climbed on a mountain with this received item. O jiva-ātmā! Tell me whom did you donate? A donation given to a worthy person and a seed sown in a fertile farm always yields fruits. Body should be properly restrained, mind should be trained by yoga, and mental defects should be purified. If you have something in little quantity, donate little; do not refuse to donate if you have something. Who sows the seed of Viṣṇu-name in his heart, it certainly multiplies infinitely.

Even if someone donated heavily, took holy-dip in all the pilgrimages, donated millions of cows there, and performed all other rituals but he cannot comprehend the secret of God since it is bound by his karmas. A pot kept upside down cannot collect water even from heavy rainfall. Who is as fortunate as King Duryodhana who saw Viṣṇu in his royal court! However, even he could not comprehend Viṣṇu and remained devoid of the reality of Viṣṇu. Those who chanted and practiced austerities also perished without obtaining the true path. Even they could not attain mokṣa, whose clothes used to be dried in the sky, i.e., the siddhas, also could not attain mokṣa due to their pride.

Nobody can be as arrogant and proud as Duryodhana who caused the MBh war. Other smaller people also cause several smaller battles every day. Only Kṛṣṇa's threefold māyā has caused the spread of the universe, none can match his māyā. Sahasrabāhu also fought fiercely, who else fought like him? The way Lakṣmaṇa shot sharp arrows, who else did? In the svayamvara of Sītā, there were famous sages and ascetics present but none could lift the bow of Śiva. They could not cross over *bhavsāgar* because of their false pride, neither can they whose clothes are dried in the sky. When I desired the liberation of 12 koti, I established my station in the middle. First Viṣṇu tested Prahlād in various ways. When he passed all the tests being endowed with fine qualities, Viṣṇu gave him Darśana in the form of Nrasingha and liberated 50 million people with him. Prahlād attained the abode of Vaikuntha with divine body and great wealth. In the Lankā war, several black, one-eyed, and three-headed evil demons were killed. Among those demons, Meghanātha first fought with Hanumān whose sound was heard beyond the ocean, he was killed at the entrance of Lankā itself. Paraśurāma also killed several kṣatriyas. Only 70 million reached heaven that were dear to Kṛṣṇa. They all attained divine bodies and great wealth. For the mokṣa-seekers, their high position is obvious. O deluded ones! Truthful Yudhiṣṭhira liberated 90 million people because he was dear to Kṛṣṇa. They all reached Vaikuntha and received divine bodies and great wealth. O mokṣa-seekers! His or her high position is obvious to everyone. I wanted to liberate 120 million people so I took incarnation that caused great trouble to evil people.

Who merely narrates the Vedas and the Purāṇas literally without understanding their meanings, is deluded, and utters nonsense. Your life reduces by day and night and your breathing is decreasing. Even sun and moon have fixed lifetimes, what to talk of you! Kāla himself is alerting you from inside yourself. Beware and save yourself. You are a mere dot of blood, o human! You slander others and face the consequences but you still do not ask for the guidance of guru.

One kind of sorrow is caused by the injury of brother Lakṣmaṇa, one kind of sorrow is caused by arrival of a young lady to the home of old man, one kind of sorrow is caused to child whose mother dies, another kind of sorrow is to take birth in a low status family. One kind of sorrow is caused by breaking up of an old relationship. O Lakṣmaṇa! Your fine qualities have neither beginning nor end and yet you could not bear the great injury of weapon. Did you have the broken bow of Śiva? O brother! Could you not understand the tricky attack of the enemy? Even 10-headed Rāvaṇa could be killed by the arrow of Lakṣmaṇa. I did not expect that you yourself would fall from the power of the enemy. Lakṣmaṇa! If anybody knows the greatness of my name, I liberate him from the worldly attachment and take him to Vaikuntha. However, without your capability, my army generals are useless and stations of three worlds are void. So what if I conquer Lankā? So what if Rāvaṇa is killed? So what if Sītā returns home? O brother full of fine qualities! What can I do? I have lost diamond in exchange of oilseed.

O Lakṣmaṇa! Did you not perform any duties? Did you spit opposite sun? Did you clean bronze pots standing up? Did you disturb the husk of a roof? Did you turn away an invited Brāhmin? Did you steal a pot from a potter's house? Did you steal green fruits from a gardener's garden? Did you break the alms-pot of a beggar? Did you break the sacred thread of a Brāhmin? Did you steal money aggressively? Did you ever separate newborn calf from its mother cow? Did you ever startle a cow drinking water or grazing grass? Did you ever kidnap somebody's wife? Did you kill your own brother? Did you attack a woman with a sword? Did you brush your teeth walking on the road? O Lakṣmaṇa! Tell me! Which of these crimes caused you to be injured by Meghanātha's attack?

I neither fell from my duty, nor spit opposite sun, nor cleaned bronze pots standing up, nor disturbed husk of a roof, nor turned away an invited Brāhmin, nor stole a pot from a potter's house, nor stole green fruits from a gardener's garden, nor broke alms-pot of a beggar, nor broke sacred thread of a Brāhmin, nor stole money aggressively, nor separated newborn calf from its mother cow, nor startled a cow drinking water or grazing grass, nor kidnapped somebody's wife, nor killed my own brother, nor attacked a woman with a sword, nor brushed my teeth walking on the road, nor did I burn a forest, nor did I beat a traveler to steal his money. Only one crime I committed. When you had gone to kill that magical deer, I disobeyed your command because Sītā blamed me without my crime. Secondly, once I laid down on your seat during our exile. Only these two crimes were committed by me.

Queen Rukmani was sent with several servants to this world. She was sent to a king with golden throne. However, even she had to return with empty hands. Mist starts falling during the night, the sun spreads its heat during the day, wind flows cold and hot, and clouds make heavy rains. Worldly people wake up with a concern for farming when water starts falling down. However, some fools do not wake up even then. The body on which I now wear woolen clothes, I used to wear soft clothes. The hands by which I now turn the rosary for chanting, I used to count diamonds. However, I left all these things for liberating 120 million people. Who can console without Rāma, Lakṣmaṇa, and Sītā? Only diamonds can match diamonds, nothing else. Problems fall even on perfect people, just as sun can also fall down in the battlefield. Sad ones will become happy and will attain profound position. Rain can cool a stove but not summer heat. Similarly even shower of knowledge cannot pacify those who are not oriented towards guru. Those who do not sleep on bed and worldly comforts and do not care for diamond necklaces can connect with God. In his adolescence, Lakṣmaṇa played with horses and other sports, but regarded his brother Rāma as guru and followed him into the forest. Lakṣmaṇa, Sītā, and Hanumān always accompanied Rāma in his happy and sad times. However, I am alone here without anybody with whom I can share my happiness and sorrow.

O creature! You have forgotten from your jealousy, you unnecessarily roar about your weak body. This weak body will perish one day and so will your kingdom. Why this unnecessary roar? Why accept grass without grains? Why utter bitter words from mouth? Being deluded you argue arrogantly. Being deluded, you selfishly kill other creatures just as wild animals do. You eat inedible due to your voracity. Those who were extremely malevolent were caught by Yama. Those who had invincible forts in Lankā were also perished. They fell down all alone without any military procession. They also were swallowed by death without any musical farewell for whom pleasant musical instruments used to be played. O Hindus and Muslims! For the benefit of your jiva-ātmā, why do not you chant God's name? Wealthy kings, pauper beggars, powerful lands, muslim leaders, queens, curly haired fakirs, crowned gurus, noble men, devas, yogis, priests, trees, their ages are decreasing every day. Who among them can escape death while path to death is same for everybody? A brute man does not change himself and calls worldly things as his own but God connects with only a truthful person. Worldly things will get scattered as if wild leaves so these are neither yours, nor mine. People are distracted. Creatures born of sweat, birds born of egg, mammals born of womb, and plants born of sprouting, all will die ultimately and fade away like dry vegetation. When body and jiva will separate, body will not be of value of even two paisa. Therefore, you should perform suitable karmas from this body. This body is not worth anything else and is insignificant. Jiva-ātmā has brought the body with itself but eventually it will go alone. It took some time to enter the world but will not even take a second exiting the world. Happiness and sorrow are obtained based on destiny. Leaves will fall off the trees and will not come back. Tender buds wither away from cold just like green vegetation. In springtime, flowers and leaves bloom again. O creature! You remain lost and your existence is unknown. One who did not chant "Viṣṇu-Viṣṇu" and did not sing the glory of God, o brother! Yama will destroy him. With the presence of jiva-ātmā, body did not earn beneficial karmas. Messengers of Yama will be painful and you will lose all your dwellings.

Gorakh was awakened by the awakening of his wisdom. He was in samādhi state for 36 eras without any desire or māyā or any other support. Gorakh is unique in this regard. Brahmā renounced his own creation and he is unique in this regard. Destiny is imprinted in everyone's forehead by God but that which is imprinted in God's own forehead is unique. All are related to each other but the relationship of Gauri and Śaṅkara is unique. There are many brothers in the world but unity of Rāma and Lakṣmaṇa is unique. Many people wear turban in this world but the turban of pride that Rāvaṇa had on his 10 heads was unique. The bashfulness of feminine character of Sītā was unique. Celebrated accomplishments of Rāma were unique. The bridge that Lakṣmaṇa built on the sea for the sake of Sītā was unique. The mission of Rāma that Hanumān accomplished was unique. The sword fighting done by Kumbhakarana and Mahirāvaṇa was unique. The kingdom that Duryodhana enjoyed was unmatched. The rāga played by Kṛṣṇa's flute cannot be compared. Traveling done by fast horses cannot be done by common asses. The groups of swans cannot be compared with that of cranes. Just as Udyāval and Vāsuki are called fine *nāgas*, snakes, will the common reptiles eaten by eagles also be called *nāgas*? Nāgar creeper is considered the best vegetable since it is the sweetest and most easily digestible; can a wild desert plant *kukar* be compared with that? Wherever evil people commit crimes, they do not succeed in great work. Kaṃsa, Keśi, Chāṇūr, Madhu, Kīchak, Hiraṇyākṣa, and Hiraṇyakaśyapa, all these demons were killed by Kṛṣṇa and Baldeva. O Mandaleeka! Why do not you see? Nobody can rule over this earth even a bit.

An evil person makes friends with pride and jealousy using five kinds of mental defects on the evil path. He is evil who has enveloped the whole world with his influence. Evil mind is the product of evil person. He hides like dirt in a black cloth. Evil people who do not follow the true path will go to hell and will never become great. Wherever evil person will spread himself, there nothing will fructify. Just as indigo spoils a cloth, evil spoils a sādhu. Just as flower has fragrance, light of God is enlightened. Benevolent acts are showered on the world just as clouds shower water on the earth. God has done such benevolent actions for the world, like producing milk in the nurturing breasts of mother.

Shukla Hansa

There are many towns in the world such as Śrigarha, and Pātan but I have chosen to incarnate in Nagore with a deep water level. I have come to conquer the unconquerable arrogant person, to imprint dharma on person without dharma, to correct the incorrigible, to discipline the undisciplined one, to humble the proud one. I have even destroyed some evil people. I have fulfilled desires of some by granting them heaven and put others in hell. I performed *homa*, decided a day and took thousands of forms to appear at Chhāpar, Neembi, Dronpur, Sudariyo, Chheela, and Balundi. Apart from these places, I have also seen Nagore area, Ranathambhore, Gāgaronagarha, Kumkoo, Kanchana, Saurāshtra, Maharashtra,

Tailangānā, and Dilligarh. I have traveled the whole world and have seen it. I travel on land and preach at Gujarat, Sapādalakṣa, Mālav, Parvat, and Māndu. I performed homa at Khurāsan and Lankagarh. I rested at Eedargarh, Ujjain, Kābul, and Sindhapuri. Why do you roar like sea? Do you do that from your own power? From which quality was it sweet and later became salty? I churned the sea with Vāsuki Nāga as leader and Sumeru Mountain as churning stick. From the things thus yielded from the sea, I killed the demons with trickery. Ten-headed Rāvaṇa had a boon so I killed him with great secrecy and trickery using special arrows to behead his 10 heads. O Rao! Do not argue with Viṣṇu. You are encouraging demon-lineage by arguing with Viṣṇu. I am I and you are you. Only great qualities mark the dynasty of true guru. Why would he frighten with a roar? He bears the fire of even the thousand-headed dragon. I do not have father, mother, sister, brother, or any other relationship. I do not have any other acquaintances also. I only have those who depend on Vaikuntha and desire mokṣa every moment. O creature! Chant "Viṣṇu-Viṣṇu" which has infinite qualities. Guru has told with several evidences that supreme element exists in thousand names, thousand places, thousand villages, musical greenery, water, sky, 14 abodes, three worlds, Jambudwipa, and seven undergrounds. He is world-creator here and there, small and big species are created just by breathing. By the māyā of Kṛṣṇa, I resemble the infinite showers from the clouds. Who knows if I am deva or of devas? I am incomprehensible *alakh*. Who knows if I am deva or human and who knows my earlier secret of previous form? Who knows I am omniscient. Who knows I am Brahma-knower? Who knows I am Brahmachāri? Who knows I am a little eater? Who knows I am man or woman? Who knows I am an arguer? Who knows I am enjoyer of tastes? Who knows I am a yogi, enjoyer, or God? Who knows I am thrifty or kind? Truthful or liar? I am myself unkind and kind, evil and noble. I destroyed the entire lineage of demons and defeated Kauravas. I killed demons in the Rāma form and with the help of forest-dwellers rescued Sītā. I am called kind one and I am destroyer. I took care of 16,000 kinds of dependent gopīs as Kanhaiyā and I am Yati. I am asceticism of ascetics. Who depended on me, always earned victory. I protect someone whom I wish just as vegetation protects a leaf in the cold winter.

Many congratulations to those princes. They deserve heavenly congratulations. However, this prince-status will perish one day. One who wanders in the world, how can he be a prince? Husk without grain is useless. Family without good karma is useless. Wealth and huge military is like smoky clouds. O creature! Chant that God by whose chanting you will practice the unique dharma. If you did not earn devotion in your good or bad health, with life force in your body, in conscious state, with your capable body, with your respirations, Yama will destroy you. For you did not sing the glory of devas, humans and God in your speech. Neither mother, nor father, nor sister, nor brother, nor friends, nor other worldly people will help you. Messengers of Yama are atrocious. They will demand the account of good and evil karmas from only one person and will not listen to any recommendation from others.

O Yamarāja! You have punished the whole world. You have not left anyone's body alive. Worldly illusion takes people to the mouth of Yama, death. None can escape that. Perishable body has no value. Ignorant are lost in its delusion. While I looked on, several devas, demons, men have perished and have gone to unknown place. Invincible warriors like Kumbhakarana and Mahirāvaṇa have also gone to such place. Lankāgarha was a complex fort once. There Rāvaṇa used to rule once with whose bed nine planets used to be tied. By whose terror, even devas were terrorized and insecure, Rāvaṇa was extremely shrewd. However, due to his lust for Sītā, he was destroyed by Yama and lost his existence. O creature! You are insignificant who thinks of working through his perishable body. Worldly people were attached to soap and other colors and were perished because all such decorative acts are futile. Worldly people are attached to such things that have no essence. Such splendor attracts everybody but one who is attached to dharma is appropriate. Fool people wander for worthless actions. He does not know that flesh and blood of his body will be wasted. Do not get attached with fake jewels and glass-stones. These are items merely for pomp and show. O creature! Why did not you in your healthy state, in your lifetime, in your capable body, while breathing chant Viṣṇu and why did you have false pride? Do you recognize the group of ferocious messengers of Yama coming to you? They will not wear anything on their heads, foot, or bodies. They neither have bows, nor arrows, nor armor. They will find you. There in the city of Yama, there will be fine silk, ghee, fine house, cold drinking water. There will not be bed, mattress, or any kind of house. Nobody will have mercy for you there and nobody will do favor for you. There will only be cruel and brutal messengers of Yama. They destroy weak and strong all kinds of bodies. They slay everybody without preparing him or her. They are cruel and destroy the evil ātmā like a powerful wrestler. After the death of a person, people of Kaliyuga cry like crows unnecessarily. Astrologers, pundits, priests, and scholars of the Vedas and the Purāṇas, if they have not understood its significance then their books are hollow. Why chant ghosts and spirits? That is pure fraud. If you are oriented toward Kṛṣṇa, you will attain divine bodies, attain mokṣa and coming and going will cease forever. One who has made resolution for tattva will certainly attain mokṣa. Why do not you chant that infinite Brahman? Regard Him as omnipresent and attain Him instantly. O creature! With focus on the foundation one is benefited just as trees sprout branches. One who ignored the foundation has spoiled both his life and death. One who is not stable on his duty day and night, cannot avoid Yama. One who focused the foundation and asked the fine brahma-tattva from guru has known the life-process. He will benefit in his lifetime and will not suffer loss after death.

Only Kṛṣṇa is true God and devotion to him is beneficial. Good karma never is wasted. Plough your fields well, implant fine seeds in it, and make protective boundary of prāṇāyāma. For the welfare of jiva, do farming and send a protector for protection. Beware of evil people wandering around; they may destroy your intellect. Be indifferent to the world, concentrate the mind, and put efforts for jiva. Do farming for jiva; do not let wind stop the growth. In the matured state of consciousness, there is neither deer, nor any other animal, nor peacock, nor any other bird, nor rats destroying the farm. O brother! Make a wise person your guru who can protect your mind from desires, illusions, and bondages. O brother! Worship the same God that King Yudhiṣṭhira did. So-called yogis are devoid of yoga, even in their shaved heads there is no wisdom. In this Kaliyuga, two people are lost, father and mother. Father is hopeful of his son ploughing his field and fetching the water from the well. Mother is hopeful that she will get a daughter-in-law and that will increase her fame. I have come here with the command from Shambhu and have established a dharma-throne in this desert. I am capable of weighing mountains on my two arms and can twist them like a mustard-seed. I pacify all the species in a second. I am an eternal yogi, incarnated and sitting in a posture. Somebody's servant asks about his future for him, a shepherd also asks similar question, folk of Kaliyuga ask similar rudimentary things. Somebody wandering on the land of sand dunes asks me about his lost goats. A hunter asks which defect of his caused his shot to miss the target. O fools! Deluded rustics! Just work hard and fill your stomach. King sitting in a palace asks, "Swāmiji! How long will I live?" A servant and a property owner also ask me the same question. A barren woman asks me, "how is my destiny? What will make me pregnant?" In the Tretāyuga, I traded diamonds. In Dwāparayuga, I grazed cows and played flute. In this Kaliyuga, I grazed goats. In the past, I reformed nine leaders of dacoits and it is the turn of tenth one named *Kālanga*. I have taken the vow of spreading the preaching in desert land and I am sitting here to gamble with people of this land. From one block, I have won nine blocks. Tell me! Will you find such a gambler elsewhere?

Who shakes his neck, vibrates it, and worships a stone idol but this is not the message of God. Look! How unjustified are ignorant people of the world! To worship stone is like guru worshipping his disciple because stone idol is made by human and then to worship that is like guru falling to the feet of his disciple. Idols made of silver, wood, and other material are covered by different kinds of clothes. People fall on the ground and bow to them. By such rituals, God can never be obtained. A donkey is better than an ignorant person who is without dharma. A dog is better than an idol. Dogs bark and guide one's way but ignorant Brahmin awaken mutual hatred. Worshipping ghosts and spirits is useless and pure fraud. Why beat husk without grain. Sesame seed without oil loses its value. O creature! Do not sow in barren land and do not build ponds in desert land that is a futile effort. Why search for water in the pond that has no water. People doing that will remain empty. Those who look like sādhus wander uselessly on this earth, lay half-naked in cemeteries, and uselessly worship stones. There is no siddha among them. O creature! Do not get deluded by them and search for your real path. A person, who has renounced the dualism and attachment with this world, will attain divine position.

Ignorant man and evil creature say that the Vedas and the Purāṇas have only created falsehood. Fire is not just fire but it is like a diamond in the ring of devas, a divine person has created it for the benefit of dharma. O creature! Illusion of Kaliyuga is to be severely criticized. Recognize the command of guru and his methodology. Pride of dharma and jāti will empty you from all sides, just as an insect harms grains. By maintaining truthfulness, evil can be destroyed just as thirst can be removed by water. I am complete person, trade knowledge jewels with me. Only they will do that whose hearts are open, the blind will not. See and obtain me, non-eater who has spread his divine play in all four sections of the earth. Chant that which is beneficial and which provides you a divine body. I kill some; liberate some, those who are without action will fall to Yama. I give pacification and save devotees from heat of different kinds of sins, this is my only tale. I omniscient have come to this desert land. I have myself spread my play. I am going to orient people toward 33 koti devas, those who came to me are contented.

Kankedā tree covered with greenery is my temple and my dwelling. If I manifest my true form, all four directions and nine islands may start vibrating. Those with fine qualities are my dedicated disciples and I am a slave to such disciples. Those with fine qualities will go to heaven and rest will be frustrated. Those with fine qualities will go to enjoyable abode, their house is Vaikuntha, and this is my message. They will get nectar-like sweet food and comforting bed that they desired. O people! Awake and look! Do not waste your life-light. One day you will also be cheated in the world. O nice farmers! I am not putting my preaching in their ears that do not repeat my words, do not listen to my preaching, do not follow my teaching, and do not practice my instructed rules. They will not meet devas in whom symptoms of Kaliyuga exist.

Enjoy bitter and sweet food and taste the sweets too. You will have to sleep on the land beneath the warm covering of sky. Whose mind is stable sleeps naturally and wakes in sacred time of dawn. O jiva! To cross *bhavsāgar* and to attain heaven, listen to my preaching.

O yogi! Know the secret of yoga. O qāzi! Recognize the message of the Qurān! For what sake do you kill cows? How did God grant you this permission? Śri Kṛṣṇa grazed cows in the forest. We have faith on such formless Kṛṣṇa that cannot be seen by eyes and cannot be described by speech. Same God is stationed in desert. O shaved sādhu! Meditate on him. Those are to be severely criticized who did wrongdoing, Allāh will ask for their account.

Keep body and mind pure, be restrained, and do not let happiness perish. As the world criticizes you, keep fulfilling your duties. Avoid attractive women just as deer runs away from the vibrating sound of a bow.

Do not be deluded by those who have forgotten themselves. The dry trunk of a tree without leaves cannot be expected to have flowers. Some drink their life force to know themselves. True guru can be easily recognized but blind people utter nonsense. Terrible karmas force you to take birth again in this world and deluded person cannot go to heaven. Finally, God will check the account of creatures but people of this world do not know this. Husk without grain, speech without essence, and family without auspicious karmas are futile. Without doing devotion by body none can go to Viṣṇu's door.

This body has millions of pores, nine apertures, and nine cavities inside them. O gardener! Why do you water this garden? This will perish one day. Always speak pleasant speech, listen to the name of Hari-Viṣṇu, why do you utter nonsense? Rarely has someone gone on true path. One who could not go on the true path, he has only beaten the husk in the field. One who has less mental orientation for Rāma, why accept such a husk without grain?

Externally nine apertures and spaces are seen. In this temporary fort, a king stays. In this fort, nobody could stay permanently. Even gurus and peers also lost their bodies.

While I sleep, if the night passes away and sun rises, then moon also gets embarrassed, sun also get embarrassed, earth and sky also get embarrassed, wind and water also get embarrassed, 18 kinds of vegetations also get embarrassed. Thousand-headed snake in seventh underground also get embarrassed. Earth sustaining bullock also gets embarrassed. Siddha, seeker, and sages also get embarrassed. Creator of the universe, God also gets embarrassed. I chant God whom 7.8 million do, who else will liberate the world?

A deceiver has devised a nice fraud. First, he broke it off by repenting on earlier sins. By that deceiving vibration, by the Vedas, by character, by sound, wind is vibrated, who recognizes that fraud? His cycle of life and death is ceased.

O Alakh! You are beyond comprehension. You are infinite. Which act of yours should be worshipped? What can this compare with?

One, who mounts on horse, does not wear turban, who can think about his acts? People of superior mind will join him but those of tough hearts will suffer in the hell. In the living state, I do not destroy anybody's karma but after death, one's fate is dependent on others. One who neither washes one's hands for cleanliness, nor washes his feet, does not deserve God Nrasingha. From infinite eras, I am the king of space.

You have shaved your head but not your mind. Deluded mind and greedy heart will destroy you. You have no compassion, you hear glory of devas but you slander and kidnap which is not befitting. If you lose guru's refuge, you will incur heavy loss. People doing wrong actions lose their entire lives. They will be crushed and wander in 980,000 births in different species and will hang upside down in hell.

Land is fertile and farmer is nice. Produce with discriminating labor. With the divine grace of guru, search inside you. Beware of thieves entering your heart. In the desert land, enlightened guru has appeared. Look out for the secret with his help. I feed all the species of three worlds instantly. Nobody can compare with King Karṇa in his charity and donations who kept his hands ready for donating gold. He donated fertile cows 21 times. Never seen a mountain like Sumeru, pond like ocean, fort like Lankā, father like Daśaratha, mother like Devaki, wife like Sītā who never had any pride, servant like Hanumān, power like Bheema, king like Rāvaṇa whose power was famous everywhere but was killed for a woman and Lankā was rehabilitated. Why do you indulge in wealth? In this Kaliyuga, Brāhmins have forgotten the Vedas, qāzis have forgotten their scripture, yogis have forgotten their yoga, "shaven-heads" have no intelligence, and parents have forgotten their duties. Father thinks when will my son plough the field and fetch water from well. Mother thinks when daughter-in-law comes to her home and musical instruments will be played for celebrations. I have established my seat as a messenger of the almighty. I weigh mountains in my arms, i.e., examine and balance the people and spread my preaching. I instantly pacify all the species properly. I am yogi established in my posture. Plough-drivers ask, shepherd asks, people of Kaliyuga ask, if they can recover their lost goat lost in the desert. Hunter asks why his arrow missed the target. O fools! You are deluded in your rustic behavior. Work hard and earn your living, do not harm animals. Emperor sitting in the palace asks, "Swāmiji! How long will I live?" Similarly, property owner asks, servant asks, and people of other tribes and jātis ask. Women also ask and barren women ask about their fate. I traded diamonds in Tretāyuga, grazed cows and played flute in Dwāparayuga, and grazed goats in Kaliyuga. I have already sent nine demons back, now it is turn of Kālanga. I have started preaching in desert land. I am a gambler and will win over them and send them to their predestined path. I will win nine blocks sitting in one block, have you seen such a gambler elsewhere?

O creature! Beware of ignorant night of the world, will you keep sleeping even being conscious? With ātmā soon to depart from your body, why do you hide mental enemies? Take the key and call the door attendant. Let the heart's lock be opened. Chant and remember the God who has been remembered by omniscient sages. One who remembers God will never be defeated. Whenever you get the opportunity, remember God. Bound by wind-bondage, this fort of body is weak. It is like water-pot. Water flows away when pot breaks. This body will be of no use. With strong faith, care for this body like a gardener cares for his garden. This body will be thrown off in fire like a bundle of sticks. Be pure, take bath, be restrained, and clean your body with pure water. According to the commands of guru, be humble and forgiving, and chant God's name with rosary. Why do not you spend your favorite thing? Why are you saving that? It is more beneficial to spend it than saving it. Home is far away, this is a temporary visit, and it is unreal. If death occurs today, it will be just another day tomorrow. If something meaningful can be done, it should be done now. Afterwards, you will be just left weeping. In your youth you did not restrain your mind, now watch its defeat.

Rarely someone walks on the path of self. They are heroes or they are slaves. When battle is announced, only enduring ones stay on. Only king struggles. Cowardly people run away. Fools do not battle.

Chant Gorakh, chant Gopāl, chant Lal, chant Gwāl, he is playful God. I have manifested myself on nine-section earth but only some know the secret of my origin.

The moon is near but the sun is far. The 900,000 stars are neither near nor far. The moon is 900,000 yojan far from the earth and the sun is 900,000 yojan far from the moon. I discuss that which is beyond all.

The 24 ghosts are magical demons. They will perform trickeries, will whirl their seats like wheels, sit on them, and establish themselves in the society. Knowingly they will delude the world. Fake person, false in the heart, will spread fictional stories. Who does not accept the commands of guru, will accept 10 mental defects in their homes. Who establishes imaginary thoughts, is a great sinner. Eventually, he will burn and perish completely. He will not ride on the ship of true guru, nor will he be dear to guru and God. He will take sand in his hands and manifest wealth with recitation of mantras. They will feed the dead wooden horse, will sit in air, and make dead man laugh. Just as sun, moon, wind, and water are established in their positions, my position is at guru-seat in Samarathal. Guru says, "O human! Do not forget this otherwise you may go to hell."

Who desires fame by fraud and deceit, is childish. Such people do not have the real wealth. My guru has taught me to purify all with real knowledge and initiate them. Whenever people have left the path of knowledge and distracted on path of ignorance, God has destroyed the sins. The elephant of attachment is tied with the death. The death is always touching human life. One who is absorbed in chanting God's name can avoid it. Body is city, mind is its king, and five vital forces are its family. Is there any brave hero who can fight with the king? Comprehend the incomprehensible Brahman, station the unstable one, walk on the unknown path. Gurus always speak the truth that fear of life and death is destroyed.

In the fort of the body, life force is supervisor, evil karma is door, and illusion is its door-lock. Having read the Vedas and the Qurān, people have only created confusions and myths have engulfed them. Siddha and seeker have common understanding; they have ascertained liberation in their lifetime. My guru has stayed in all the eras, came down to me, and called me siddha. I am natural-bather, intrinsically pure, and omniscient. Your good karmas will not be wasted. I say something and people understand something else. Without understanding my preaching, they will not be able to attain siddhi.

In the beginning, there was unbroken word and voice. There was water in the 14 abodes, an egg arose from the water, and Brahmā, Indra, and Murāri arose from that.

God has thousands of names. I arose from Ādi Murāri. That time I existed as a universal power without any support in foggy smoke. I have neither mother, nor father. I have improved my body myself. Void continued for 36 eras and universe was created in Satayuga. Brahmā, Indra, etc. were established including the whole universe and many times Indra and others were given power. The moon and the sun, both were established in their manifest form. Life force is dependent on God. That time I took the fish-incarnation and preached Satyavrata. Then I took the turtle-incarnation to help devas. Then I took boar-incarnation and rescued the earth on my jaw. I killed Hiraṇyakaśipu in my Nrasingha-incarnation and Prahlād stayed in my refuge. Taking Vāmana-incarnation, I inspired King Bali to donate and measure the entire earth in three steps. Becoming Paraśurāma I conquered kṣatriyas and did not leave even kṣatriya fetus. Among other proud kings, I married Sītā in my Rāma-incarnation and wore crown. As Kṛṣṇa, I played flute, grazed cows, overcame Kāliya Nāga, and killed demons. In Buddha-incarnation, I killed Gayāsura, rendered him useless, started a new religion, and showed people a new way. I have won nine times. I Jambharāja am supreme God. Nobody can grasp the secret of Jambheśvara and Guru Gorakh. Qāzi, mullāh, and educated pundits who criticize are fools. If you want to avoid hell and go to heaven, follow my commands. You will dwell in Vaikuntha and achieve liberation.

O creature! Arguments are to be condemned. Leave the obsessions and fascinations of mind. Whose mind is engulfed with darkness? Whose mind is shining with sun of knowledge? Who are without guru have darkness in their hearts. People oriented toward guru witness the dawn of sun of knowledge. In the old age, legs started shaking, eyes became weak, body became frail, was insulted by children and siblings. Oil cakes devoid of oil are suitable only for animals. Similarly, swan of ātmā flies off from body and leaves it useless. In the next world, God will demand your account, "Tell jiva! What kind of karmas did you earn?" Jiva could not see anything on his review and started repenting for his karmas.

Listen, intelligent! Listen, virtuous! My manufacturing is similar to blacksmith. Blacksmith heats iron in furnace and makes it useful. Goldsmith heats gold in fire and makes jewelry. Similarly, my strong will is my chisel, my preaching is my hammer, and bodily heat is like fire. If you accept the preaching of guru, you will cross over *bhavsāgar*. Leave the worldly seat and establish in the seat of ultimate bliss. For the welfare of jivas, I am like Jambharāja blacksmith.

O creature! Chant "Viṣṇu-Viṣṇu" so that brother, your mind stabilizes. During the day, you forgot God but you did not become alert even at night. What desire you have in your mind that you are still sleeping. Your body is unreal but you have a lot of attachment. O brother! What hope you have in your mind? Keep working with hands and remember Viṣṇu in your heart. You forgot God and did not follow his commands but chanted Maya in this confusing world. That creature is to be condemned who loves stone but without guru he will not be liberated. Devotee Prahlād practiced intense devotion of God and therefore he could liberate 50 million creatures. True to his vows, truthful Harishchandra sold his wife and son in the market. With the power of his donation and charity, he took 70 million creatures to heaven. Thanks to mother Kunti, whose truth-speaking son Yudhiṣṭhira liberated 90 million creatures. Since I promised devotee Prahlād, I have come to invoke 120 million creatures for mokṣa. Who is whose wife? What is dear to whom? Who is whose brother or sister? Deluded creatures of the world are dying and going. They have not recognized of devas Viṣṇu. Iron is harder than stone. Fools can mark anything for devotion.

Guru can instantly cause heat, calm, wind, water, and cloud. Kṛṣṇa can even turn desert into pond instantly. O self-deluded creature! Remember Viṣṇu so that death brought by Yama can be avoided. Stone can be made wet externally but water cannot enter inside it. Only he has known the way of living that has killed his ego being alive. If somebody comes angrily, we should become like water. Who has humility, forgiveness, auspicious actions, tolerance, pure body, watch carefully, he has arose his Ātmā into sky. Do not regard this temple and monastery as real, chant the supreme God. Consider God with you in infinite forms, make his acquaintance favorable to you, and make efforts. Seekers of mokṣa should regard creatures born of sweat, birds born of egg, mammals born of womb, and plants born of sprouting, all of them as God.

This is correct and true. I am not lying. I am near you. I cannot stay far from you. I am always contented and sustainer of truth and have left praise or insult. Control the wind, semen, your body, and apertures of your body. The tenth door is locked. Only an expert can unlock it. One who unlocks it will make a way inside it. O brother! Worship the same God that Yudhiṣṭhira did. That guru who is celibate, whose concentration on God is undisturbed, whose life force is stuck on the tenth entrance that has renounced the embrace and broken the illusion.

Consider wealth, property, carts, and other means and resources unreal. These are mere decoration items. O emperor! Do not get lost in this confusing web; be detached from such illusionary path. Collecting elephants of 900,000 rupees, enclosing them, donating the enclosed elephants, and finally getting proud of the donation – these are all illusionary imagination. It is also sinful and unreal to please ghosts and spirits using tantras and mantras and to eat inedible food. O man! God is not obtained by taking favor from somebody, by deceit or by depending on something. Those unfortunate ones seek God from these things.

Who recognizes true guru, at his place it is always new moon and *Sankranti.* Nine planets also sit in a line and planetary position is always favorable. Sacred Ganga always flows in full force also. He has established his seat on the sacred Ganga and sacred abode. If he learns to wash, then his house will not have any unclean clothes. He applies soap with focused mind and restrained intellect and customer enjoys wearing such clean clothes. Up and down knowledge is everywhere. There is nothing else present. There is oil in sesame seed and fragrance in flower, similarly God has illuminated all the five elements. There is a motion in the spark of knowledge that stays in void. He neither sings, nor seeks fame, nor delays in going to heaven. True guru makes us aware of Brahman Tattva that makes us eternal and free of cycle of life and death.

Chant "Viṣṇu-Viṣṇu" and make your defects weak to benefit spiritually and to purify from sins. Leave aside other things and chant God. By destroying other feelings, you will attain spiritual bliss; visualize Hari in people and gods. When your breathings will lose their worldly desires, you will attain mokṣa.

Ignore the bad qualities of others, bad words of others, practice forgiveness and austerities, donate according to your capacity, and do not refuse if you possess certain things. All the three worlds are witness of the grace of Kṛṣṇa.

His grace gives immense fruits. Know the philosophical meaning of the word "Viṣṇu." Who sows "Viṣṇu" seed, gets infinite times more benefit.

I consider donation of gold nothing, donation of clothes nothing, donation of ghee nothing, donation of silk and sacred yellow clothes nothing, donation of 500,000 horses nothing, donation of elephant nothing, donation of daughter (in marriage) nothing. I regard purity and bathing very highly.

From unmanifested Niranjan has arisen Shambhu spontaneously. Brhaman evolved there was neither sun nor moon at that time. There was neither wind, nor water, nor earth, nor space. That time there was neither month, nor year, nor moment, nor hour. There was no sunlight, no shadow, no hot, no cold, no three worlds, no stars, no clouds, no rain. That time there was no planetary positions, no calendar, and no fourteenth day of the month, no full moon, or new moons. There were no oceans, no mountains, no Himalaya, and no Sumeru. There were no shops, no paths, no forts, no towns, no commerce, no things, and no profits. There were no powerful kings, no sultāns, no diwāns, no sects of Hindus and Muslims, no work, no farming, no yoga, and no darśana. There were neither pilgrims, nor mosque-dwellers, no chanters, no ascetics, no mares. There were no brave heroes, no sword, no kṣatriyas, no battles, no wars, no lions, no cubs, no birds, no swans, no peacocks, and no pigs. There were no colors, no tastes, no clothes, no greasy substances, no wheat, no rice, and other edible substances. Neither mother, nor father, nor siblings, nor children. No breathing, no word, no conscious jiva, no body, no man, no woman. No sin, no merit, no noble woman, no characterless woman, no compassion. Shambhu arose with māyā spontaneously from abstract Niranjan. Brahman is born involuntarily. O emperor! Listen to this statement.

O qāzi! O mullāh! Listen! Man, Woman! Listen! I am the only foodless person who has preached you to follow the dharma-path. If you stop killing innocent animals with sword, then your reading Kalmā and remembering Khudā is meaningful. Only he has earned entry into heaven whose heart is ready to sacrifice for the truth.

I have laid down the bed of natural character and remained indifferent. Whom should I narrate this story of several eras and births. When it dawns, the light spread in the entire world and owl turns blind. People believed that they found the true guru and he removed all the delusions and showed the true path. Whoever recognized the true guru, got the evidence, merged with the ultimate, and fulfilled his desires. Who did not recognize the guru, did not get the way, was tied with several bondages of delusions.

Follow the yoga-path of hāli and Pali. Pāli has tilled the field. Make the sun and the moon as your bullocks and the rivers Ganga and Yamuna as your reigns. Sow the seeds of truth and contentment; your crop will touch the sky. Consciousness is at the entrance, deer will not be able to eat away the crop. With the grace of guru, by omniscience, by experience of brahma, and by natural bath, you will attain wealth and success in your home.

Look! Mind often accepts the easy. It forgets service but remembers bathing. Look! Mind is full of delusions and is empty in conscience, and looks wet externally. Look! Mind accepts only illusions and gets happy by them, having removed God from heart; it has united with the evil. Look! Deluded mind call tree of āka as tree of nut. You accept only delusions repeatedly! Do not be forgetful and do not be deluded. Why do you desire flowers from the dry wood having no leaves?

She was the queen of Mathurā and head queen. She robbed the pilgrims and their horses. Tax collectors chased them, robbed diamonds, and precious jewels. Due to not fulfilling the vows and promises, she serves as an animal at the well. O creature! Why do you blame Viṣṇu for that? You designed your fate to bear the punishments.

Where rough clothes are worn, desert fruit is eaten, cows are domesticated, farms do not have boundaries, water is drunk from deep wells, devas and men are locked in prisons, I have incarnated in such a country. I will test the innocents, select them, and inspire them for mokṣa. I had made promises to Prahlād so I will not embarrass both the guru and his disciple. I identified the clan of people and united them with 33 koti devas.

Who break the panth-rules, slander the guru, and have enmity with him, they are misguided, atheist, and unworthy, they are violent and killers of animals. They are the messengers of evil and are similar to ghosts and devils. They are demons and eaters of wrong food. Not by birth but by their karma, they are chāndāls. Who nourished himself by killing other animals; his ātmā will be caught and put in the dark hell. He will be beaten by hunters, will be examined, and tortured and none will rescue his screams for help.

Who has faith on God is a true Muslim, this was told by Muhammad, and it is not hidden. Read namāz inside your heart, this is your duty and you will know God completely. God will grant mokṣa to those Muslims who renounce their falsehood and have the right faith. You should get this information through mullāhs.

O Jāts! Listen! Message of devas has come to you. Why do you walk in the dark when light is present. Have you forgotten the way of the guru? Why do you keep your mind impure when the pure water is present? Why are you spoiling your earning? Why are you throwing your diamonds in hand and catching fake pieces of glass? O Jāts! Listen! Message of devas has come for you.

I have adopted the clothes of poverty on my body. Watch my doable actions. Know my behavior, then bow to me, and do not forget my preaching. Only water is fetched from the rivers but diamonds are also obtained from the ocean. Wind flows through all the creatures by God. In the evening, you should practice devotion toward God; where is that profound water reservoir? Where is that water which is waved by God and merges into God? I reign in the void and awaken the sleeping creatures.

O yogi! Why move your deerskin and sandals? O yogi! If I desire, I can stop even the rising sun. If I desire, I can make Sumeru and Udaygiri collide with each other. If I think about queen Rukmani, I can populate three abodes here. I can make 900 rivers and 89 streams flow in this desert. I fought with Rāvaṇa in such a way that having broken Lankā, I rescued Sītā. O yogi! I can bring my arrows here by which I killed Rāvaṇa. I can bring Hastināpur here if I so desire. If you move a golden deer, I can shower the stones by my imagination. O yogi! Why do you show off your deerskin and sandals?

Just as a dog entangles his head in a pot as soon as he spots a piece of bread in it, you also shaved your head and became ascetic without understanding yoga-secret. Without introduction, both guru and disciple do not attain mokṣa after death.

God has incarnated on the earth from the heaven. Tell me, why has he come? He is one, foodless, and is established as pure light manifested. He promised to Prahlād about the welfare of 120 million people. Even one person left out of 120 million will embarrass both the guru and the disciple.

O creature! Chant "Viṣṇu-Viṣṇu," you will be benefited just as you save every coin to accumulate millions of rupees. Your body will become divine, you will stay in Vaikuntha, and your fear of life and death will cease.

O creature! For the welfare of your jiva, chant "Viṣṇu-Viṣṇu" repeatedly. Your age is decreasing every moment and the death is coming closer to you. You have changed from youthfulness to old age, then why do not you awake? Death will certainly destroy you. O fool! Why do not you get oriented toward guru and ride on the boat? Being self-oriented why do you carry unnecessary burden? As you remain embarrassed with the world, you will keep sinking down.

Appendix B
Hindu Myths in Jambheśvara's Śabdas

His statements mention Kṛṣṇa at a number of places, "Without the grace of Kṛṣṇa, water has never stayed and will not stay in the water pot. But His wish can turn the impossible into possible."[1] Similarly, śabda-14 is completely dedicated to Kṛṣṇa.[2] In śabda-16, he says, "O Jats! I have preached to you. This is evidence of the grace of Kṛṣṇa." In several of his sayings, Viṣṇu also is mentioned, "O Person! Only that Viṣṇu will help you sail your boat like a boatman Kevat. So just meditate on Him."[3] In śabda-70, he says, "Kṛṣṇa is the final truth and one who does good deeds for Kṛṣṇa succeeds in his efforts."[4] In śabda-33, he glorifies Kṛṣṇa's BhG, "The BhG is not just a poem to be sung in music but it is a color that brings back colors in a broken piece, i.e., it removes the ignorance."[5]

Similarly, several of his sayings show glimpses of the Rāmāyaṇa and the MBh. In śabda-85, he says, "We have not seen a father like Daśaratha, a mother like Devkī, and a wife like Sītā. None of them had any sense of pride in them. We have not seen a devotee like Hanumān, a powerful man like Bhīma, and a king like Rāvaṇa."[6] He was quite impressed with the mutual love among Rāma, Sītā, and Lakṣmaṇa. In śabda-63, he says, "Lakṣmaṇa, the brother of Rāma, used to play with horses in adolescence. However, he accompanied his brother in the exile. Lakṣmaṇa, Sītā, and Hanumān always accompanied Rāma."[7] Further, he describes the agony of Rāma when Lakṣmaṇa faints away in the Lanka war, "I may achieve the victory over Lanka killing Rāvaṇa and get Sītā back. However, these are all useless without my dear brother Lakṣmaṇa of great qualities. Lakṣmaṇa is like a diamond comparing all the kingdom and wealth as a waste."[8] Further, he describes Rāma's agony, "Rāma was as shocked by Lakṣmaṇa's unconsciousness in the battlefield as an old man is unhappy marrying a young lady or a child cries at the death of his mother, or a poor man feels among rich people."[9] In śabda 61 and 62, he speculates the reasons for Lakṣmaṇa's injury and finds that he was devoid of any of the usual sins that lead to misery. He was also impressed

1 Jambha-BhG, p. 50.
2 Śabda-vāṇī darśana, p. 34.
3 Śabda-vāṇī darśana, p. 56.
4 Jambha-BhG, p. 278.
5 Jambha-BhG, p. 148.
6 Śabda-vāṇī jambha sāgara, p. 235.
7 Śabda-vāṇī jambha sāgara, p. 166.
8 Śabda-vāṇī jambha sāgara, p. 157.
9 Śabda-vāṇī jambha sāgara, p. 156.

with the power and celibacy of Lakṣmaṇa. In śabda-48, he says, "There is nobody as powerful and celibate as Lakṣmaṇa, one who battled with Meghanatha and Rāvaṇa." He was also impressed with Vibhīṣaṇa who helps Rāma against his brother Rāvaṇa. In śabda-27, he describes both Vibhīṣaṇa and Rāvaṇa, "Rāvaṇa's brother Vibhīṣaṇa reveals the entire secret of Lanka and Rāvaṇa to Rāma and thus acts in accordance with the dharma. Even a scholar and powerful king like Rāvaṇa had to face the outcomes of his evil deed. All his ten heads were pierced with the arrows of Rāma."[10] Praising the glory of Rāma, in śabda-32, he says, "Karṇa, Dadhichi, Shivi, and Bali all these great kings enjoyed the fruits of their great sacrifices and donations but could not become Rāma. One can become Rāma by fulfilling one's duty in accordance with the dharma."

He is equally impressed with the MBh. Praising the great qualities of Yudhiṣṭhira, in śabda-58, he says, "Yudhiṣṭhira, being affectionate of Kṛṣṇa, inspired even the stupid, the wicked and the atheist to strive for liberation and heaven. Ninety million people were directed towards heaven by Yudhiṣṭhira."[11] Appreciating the "Pāhal," the vow taken by Kunti, he says, "The mother Kunti obtained the boon of five sons for her vow and this vow enabled ninety million others to attain liberation with Yudhiṣṭhira."[12] In śabda-30, he praises Karṇa and says, "There has never been a donor like Karṇa who used to donate fifty kilogram of gold every day. He had also donated a healthy cow to a sage. He even donated his diamond tooth at the battlefield when asked by Kṛṣṇa for it. Nobody else has ever donated so much gold as Karṇa and hence he is famous as a great generous king."[13] In śabda-29, he continues on Karṇa, "An unmarried young girl worshipped the god Sun and gave birth to Karṇa. This was kept as a secret forever."[14] He is also impressed with the great devotee Vidura. In śabda-30, he says, "Vidura meditated upon Kṛṣṇa, obtained great qualities such as contentment, humility, and nobility, and thus achieved the liberation."[15] He did not like the false pride of Duryodhana. In śabda-58, he says, "Nobody will be more egoistic than Duryodhana. It was his false pride which led to the greatest war of the MBh in which countless people lost their lives and kingdoms."[16] Continuing further, he says, "Duryodhana was so arrogant that he did not honor even Kṛṣṇa, the god incarnate. It was this arrogance which led to his downfall."[17]

Many of his other statements mentions references from other Hindu epics and myths. In his Pāhal mantra, he says, "Nine incarnations of Viṣṇu are

---

[10] Śabda-vāṇī darśana, p. 123.

[11] Śabda-vāṇī jambha sāgara, p. 153.

[12] Śabda-vāṇī jambha sāgara, p. 338.

[13] Śabda-vāṇī jambha sāgara, p. 99.

[14] Śabda-vāṇī jambha sāgara, p. 92.

[15] Śabda-vāṇī jambha sāgara, p. 100.

[16] Śabda-vāṇī jambha sāgara, p. 150.

[17] Śabda-vāṇī darśana, p. 118.

adorable."[18] In śabda-58, he continues, "The Narasiṃha incarnation of Viṣṇu is the best among men, most courageous among the braves, king among the men, and of the gods."[19] In śabda-67, he says, "In his incarnation as Rāma, he killed the demons."[20] In śabda-85, he says, "Viṣṇu in his Kṛṣṇa incarnation in the Treta Yuga grazed the cows and established the social norms in his Rāma incarnation in the Dwāpara Yuga."[21] In śabda-94, he says, "Viṣṇu in his Fish incarnation saved the king Satyavrata from the huge deluge. He protected the earth in the Boar incarnation. He killed Hiraṇyakaśipu in the Narasiṃha incarnation. He reformed king Bali in the Dwarf incarnation. He killed the arrogant kṣatriyas in the Paraśurāma incarnation. He married Sītā in the Rāma incarnation. He grazed the cows, played the flute, and killed the demons in the Kṛṣṇa incarnation. He killed Gayāsura in the Buddha incarnation."[22] In line with Hindu tradition, he also regards the sky, fire, water, and the earth as gods.[23] In śabda-54, he adores the sun as a god.[24] In śabda-94, he adores Indra, Brahma, sun, moon, wind, earth, water, and other 330 million gods.[25] In śabda-101, he also affirms the existence of nine planets, a traditional Vaiṣṇava belief.[26] In śabda-2, he affirms non-dualism and proclaims that ultimate God is only one who is beyond any names or qualities.[27] With form or formless, He stays in infinite creatures forever.[28] He was a strong exponent of daily Vaiṣṇava/Hindu rituals. For all his followers, including non-Brahmins, he prescribed three-time Sandhyā, evening Āratī, singing Bhajans, performing Yajñas, meditating upon Viṣṇu.[29] In śabda-7, he says, "The day when one does not perform his daily rituals, one's mythic cow Kapila is lost, i.e., peace of that home is lost."[30] In śabda-31, he says, "A person who does not chant the name of Viṣṇu cannot attain the liberation from the cycle of life and death."[31] In śabda-86, he says, "Just chant and meditate upon the God, only He is the of all the beings."[32] In śabda-97, he says, "O brother! If your heart accepts it, just chant Viṣṇu."[33] In śabda-58, he says, "People with good actions go to heaven where

18 Śabda-vāṇī jambha sāgara, p. 30.
19 Śabda-vāṇī jambha sāgara, p. 151.
20 Śabda-vāṇī jambha sāgara, p. 188.
21 Śabda-vāṇī jambha sāgara, p. 239.
22 Śabda-vāṇī jambha sāgara, pp. 263, 264, 265.
23 Jambha-BhG, p. 8.
24 Śabda-vāṇī jambha sāgara, p. 142.
25 Śabda-vāṇī jambha sāgara, p. 263.
26 Śabda-vāṇī jambha sāgara, p. 287.
27 Śabda-vāṇī jambha sāgara, p. 31.
28 Smārikā, p. 84.
29 Smārikā, p. 83.
30 Śabda-vāṇī jambha sāgara, p. 35.
31 Śabda-vāṇī jambha sāgara, p. 102.
32 Śabda-vāṇī jambha sāgara, p. 242.
33 Śabda-vāṇī jambha sāgara, p. 274.

they get divine bodies and infinite wealth."[34] In śabda-77, he says, "People with evil actions go to hell where they are eternally bound to the cycle of life and death."[35] In śabda-96, he says, "When the soul leaves the body, one is asked for one's account of good and evil actions and one is rewarded accordingly."[36] In śabda-33, he affirms an oft-repeated Vaiṣṇava/Hindu thought, "Who has not left this world and who will stay here forever?"[37] In śabda-86, he says, "Why do you accumulate wealth and other things?"[38] In śabda-100, he preaches to a king, "O king! Your wealth, elephants, horses, chariots, or military are all futile. These are all just illusionary things so avoid them. It is better to rule them than be ruled by them."[39] In śabda-1, he defines a true guru as someone who, "has created and knows the six darśanas (Hindu philosophical systems) in which God and living world are described. A person with determination, pleasant nature, and listener of God's voice, knower of spiritual knowledge and the Vedas and a great orator is the true guru."[40] In śabda-14, he explains his preaching, "I have not written any book because spiritual experience cannot be expressed but my words are based on the message of the Vedas and other scriptures." In śabda-27, he says, "Worldly and spiritual knowledge can be learned only by studying by the Vedas. One that could not attain omniscience even from studying the Vedas is really an unfortunate person." In śabda-59, he says, "One that merely read the Vedas and other texts but did not practice their teachings, is deluded." In śabda-85, he criticizes the contemporary corruptions of different communities and says, "Brahmins have forgotten the Vedas, Muslim Qāzis have forgotten their Qurān, Yogis have forgotten their Yogic practices, and intellectuals have lost their intelligence." In śabda-92, he continues, "They study the Vedas and the Qurān but spread false stories around not the true wisdom." Further, in śabda-72, "Instead of understanding the essence of the Vedas and the Qurān, even the scholars misinterpret them and corrupt themselves."

Thus, we find a strong influence of Indian traditional myths, epics, and texts in Jambheśvara's teachings. Continuing the traditional Indian cultural glimpses in Bishnoi community, I now describe about the Rāmāyaṇa written by a Bishnoi poet Mehojī (Maheshwari 1984). Mehojī's Rāmāyaṇa is based on the main story of Vālmīki's Rāmāyaṇa that he wrote in Vikram Samvat 1575 about 58 years before Tulsidas wrote Rāmacharitamānasa. There is no other writing available by Mehojī except his Rāmāyaṇa. Some of the unique features of Mehojī's Rāmāyaṇa:

---

34 Śabda-vāṇī jambha sāgara, p. 156.
35 Śabda-vāṇī jambha sāgara, p. 220.
36 Śabda-vāṇī jambha sāgara, p. 270.
37 Śabda-vāṇī jambha sāgara, p. 105.
38 Śabda-vāṇī jambha sāgara, p. 244.
39 Śabda-vāṇī jambha sāgara, p. 286.
40 Śabda-vāṇī jambha sāgara, p. 17.

1. When Daśaratha falls ill, his wife Kaikeyi attends to him with complete devotion warranted of a traditional Indian wife. As a reward for her service, Daśaratha grants her two boons. This is different from Vālmīki Rāmāyaṇa's tale in which Kaikeyi protects Daśaratha in the battlefield and later gets two boons granted by her husband.
2. At the time of exile of Rāma and Lakṣmaṇa, their brother Bharata is present in Ayodhyā and he bids them farewell at the border of Ayodhyā city. This is different from popular tale in which Bharata is at his maternal grandfather's home at this point.
3. Rāma marries Sītā on his way to exile in the forest. In the popular version, the wedding takes place on his return from the hermitage of Vishwāmitra much before the time of their exile.
4. Rāvaṇa's soldier Bhoja narrates the great beauty of Sītā to Rāvaṇa and advises him to kidnap her. In the Vālmīki's story, Akampan narrates to Rāvaṇa the episode of Rāma killing the demon Khara.
5. In his search for Sītā, Hanumān jumps beyond the ocean and Lankā and has to return to Lankā since he covers much more distance than needed. An old woman helps him return to Lankā.
6. With the permission of Sītā, Hanumān tries to satisfy his hunger with the fruits of Aśoka Vātika. However, Sītā forbids him from doing so and instructs him to eat only the fallen fruits.
7. Hanumān himself hints about his death to the demons by suggesting that they burn his tail.
8. Rāvaṇa's sister Vārāhi asks a passerby about Lanka, her paternal hometown, *Pihar.*
9. A doctor tells Rāma that the person bringing the herb should return before dawn. This matches with Vālmīki Rāmāyaṇa's description and Hanumān is directly instructed in such way.
10. Mahārāvaṇa plays trickery and takes Rāma and Lakṣmaṇa to *pātalaloka.* He intends to sacrifice both to goddess Mālādevī. Hanumān reaches pātalaloka via seaway, asks a gatekeeper about whereabouts of Rāma and Lakṣmaṇa, and reaches the temple of Mālādevī through a tunnel.
11. When Lakṣmaṇa intends to kill Rāvaṇa, he sends his ministers to ask for mercy to Lakṣmaṇa.
12. Eventually, Lakṣmaṇa kills Rāvaṇa, not Rāma. (This matches with the Jain Rāmāyaṇa.)
13. Ecology of Rajasthan is evident in many places throughout the text. The places that Bhoj narrates to Rāvaṇa are those found in the desert of Rajasthan. At another place, Gauri Pūjā is mentioned, which is a fortnight-long festival of women in Rajasthan. Sītā performs this ritual for seeking a suitable husband. Elsewhere, Rāma and Lakṣmaṇa are described wearing traditional colorful turbans of Rajasthan. In their exile days in the forest, Rāma and Lakṣmaṇa are described constructing the water-tanks. Sītā is described fetching the water like *paṇihārī*, Rajasthani woman carrying

the water, and watering the Rajasthani plants such as kevadā, maruā, and champā. Several other medieval Rajasthani cultural glimpses are present throughout the text. For instance, Sītā asks Lakṣmaṇa for the golden deer instead of directly asking Rāma, as is the custom in Rajasthan even today (women ask their young brother-in-law or sister-in-law to pass on the message to the head of the family).

14. Influence of Jambheśvara's teaching is evident at several places. For instance, Rāma regrets killing the deer. Just as Jambheśvara had asked his followers to take a bath every day, Mehojī's Rāmāyaṇa mentions his characters taking a bath at several places, including Mahārāvaṇa in verse.

Appendix C

Bishnoi Saṃskāras ("Rites of Passage")[1]

Gotrachāra

All Bishnoi rituals and rites of passages are initiated and performed in the presence of fire. In 1485, Guru Jambheśvara established Bishnoi Pantha with the performance of a havan, fire ceremony. Today, Bishnois follow the same havan tradition in their daily rituals and in their rites of passages. Gotrachāra includes verses taken from various śabdas by Jambheśvara.

Revered *Agnideva*, fire-deity, present in everybody, is a reservoir of qualities, father of sky, mouth of five undergrounds, omnipresent, protector of all, fine and benevolent, and mouth of Śiva.

Pārvatī asks, "In which month, in which fortnight, on what date, on what day, in what astrological position, was fire-deity born?"

Śiva answers, "Agnideva was born in *Āshādha* month, in the *Kṛṣṇa pakṣa*, at midnight, in *Meen* and *Rohini* astrological positions, on the fourteenth day, on Saturday, with its head pointing upwards, and looking downwards."

Pārvatī asks, "Who is his mother, who is his father, what is its *gotra* (last name), how many of his tongues are manifested?"

Śiva answers, "His mother is Arani, his father is Varuṇa, his gotra is Śhāndilya, and Vanaspati (vegetation) is his son. He is purifier and sustainer of all wealth." It has four thorns, three limbs, two heads, and seven hands. Tied from three sides, it has entered the humans making heavy sound like a bull.

Entire universe is in its belly. It has 12 eyes and seven tongues.

The names of its seven tongues are, Kāli, Karāli, Manojvā, Sulohitā, Sudhroomvarṇa, Sfulingini, and Viśvarupa. These are fiercely blazing.

It has seven kinds of food, ghee, barley, sesame, yogurt, *khīr* (sweetened milk with rice), sweetened yogurt, and sweets.

Seven tongues are illumined by these seven foods that help humans from upside, downside, and from front. They insert the ghee, sweets, etc. into the mouth of Viṣṇu and satisfy all devas Brahmā, Viṣṇu, and Maheśa etc.

I invoke that Agnideva who is priest of yajña and sustainer of yajña and all kinds of wealth.

Agnideva is adorable by ancient sages and revered by new sages. He summons devas here.

---

[1] This information is based on my translation of a Hindi book of Bishnoi Sanskāras by Shrikrishna Bishnoi.

Through Agni, a devotee attains such prosperity that increases day by day, attains name and fame, and remains powerful.

O Agni! Yajña surrounded by you reaches devas.

The invoking Agni with divine resolution is existent and endowed with many kinds of shrutis. Agnideva should come here with other devas.

O Agni! You are born for doing benevolence to others, that is the truth and it is yours.

O Agni! We approach you every day and every night, bowing to you with our intellects.

We approach that Agnideva who is sustainer of the yajña, holder of the truth, and grows every day.

Therefore O Agni! Be accessible for us just as a father is accessible to his son. Surround us for our benefit.

Kalaśa-Pūjā Mantra

Kalaśa Pūjā is for *pancha-mahābhutas*, five great elements. The clay-pot at the havan-altar represents the earth element, water in the pot represents the water element, fire of the havan represents the fire element, and all pervading space and wind represent the other two elements. This ritual is performed at all the rites of passage such as births, weddings, and deaths. First, a plain clay-pot is filled with pure filtered water and is placed on sand or grains near the havan-altar at a prescribed time. Two copper coins are kept in the pot and its mouth is covered with a new piece of cotton-cloth. Then the havan is performed methodically with chanting of the śabdas of Jambheśvara. Later on, an elderly man puts his joined hands on the Kalaśa and with the "ādeśa" śabda of Jambheśvara; the priest chants this mantra:

> This world is full of knowledge. Knowledge is truth and truth is God. Evolved from the knowledge, the world eventually dissolves into it. The original and omnipresent God suddenly had the form of the universe in his mind. Out of his desire, he decided to multiply into many and this started the wheel of creation. He invoked the *pancha-mahābhutas* and they agreed to assist him. In this way, God created the universe with the help of *pancha-mahābhutas*. The imagination and speculation of God is infinite which created first the seven continents and nine blocks. He combined the five elements and created a solid oval shape. This developed and eventually became the earth. Water arose in the middle of the earth due to chemical reaction and Viṣṇu was born there. Then a lotus developed in the navel point of Viṣṇu. Ever-growing Brahmā-seed was established in this lotus. In this way, for the first time, Brahman was created which performs the acts of creation, preservation, and destruction by its three divine forms, Brahmā, Viṣṇu, and Maheśa. From this earth to heaven, Brahmā is the only creator of the entire creation. Jiva forgets its creator and becomes conscious

only about its body. Such people live only at the physical level. However, the Brahman is eternal, formless, and is beyond all activities which neither had any parents, siblings or family, nor will it have any attachment in future. One who meditates on formless Brahman and devotes on it, his karmas from all past lives are purified. In the beginning of time, Kalaśa was established for the first time when it arose in the form of lotus from the navel point of Viṣṇu. Then the cosmic person kept it in front of him and all devas witnessed it. God established it and this is similarly done by priests from several millennia. The priests wish that just as Prahlād helped reestablish dharma by establishing the Kalaśa, the ritual Kalaśa that they establish should also have similar effect. In addition, priests ask for well-being of everybody. Similarly, the truthful king Harishchandra had also established a similar Kalaśa with his wife and son. Priests wish that the devotees might practice dharma just as Harishchandra did by establishing his Kalaśa. Similarly, they wish to follow the example of dharma-follower Pāndavas.

Pāhal Mantra

In 1485, Guru Jambheśvara had initiated his followers into Bishnoi Pantha using the Pāhal Mantra. The water spelled by this mantra is called Pāhal. One who drinks this water vows to follow the 29 principles of Bishnoi Pantha. In this way, anyone can become a Bishnoi by this ritual. Like the other mantras, this mantra is also recited at every major rite of passage in Bishnoi community, e.g., at births, weddings, and deaths. Bishnois gratefully recite this mantra and rededicate themselves to follow the 29 principles laid down by Jambheśvara. They also recite this at the Holi festival and pay their homage to Prahlāda. Pāhal is derived from the Rajasthani word Pāla that means boundary, e.g., boundary for water-storage. This mantra given by Guru Jambheśvara is as follows:

> O devotees! First, you meditate upon Brahmā, Viṣṇu, Maheśa, and the unified form of three gods, the Omkār, bow to it because it is most auspicious for you. It provides Mokṣa to even those creatures that are not suitable for Mokṣa. This original word Omkār is the foundation of all dharmas. Both the name and the sight of saintly people are auspicious. All the sins of that person are destroyed with whom noble people are pleased and have become favorable. Jiva gets human body after reincarnation in 8.4 million species. Having attained the human body, if someone takes the refuge in true guru, one can attain the Mokṣa. Identification of true guru is that he is ever content and lover of purity. He is never indulged in selfishness and is always busy in care and welfare of others. One who spends entire life for others definitely gets the Mokṣa and gets freedom from the cycle of life and death. Therefore o devotees! Noble people advise that one should take refuge in the true guru because only by that

refuge can one live at spiritual level rising from physical level. Then one can know and identify oneself.

O devotees! By taking refuge in Supreme Viṣṇu, great sages and siddhas have also accomplished their missions. Prahlād also attained Mokṣa by it and liberated 330 million other creatures. O brothers! Truthful Harishchandra also attained Mokṣa by it. The mother of Pāndavas, Kunti, also attained success only when she took refuge in this mantra. By god's grace, she met with great noble people and finally attained the Mokṣa. She reformed both her life and death by taking refuge in Viṣṇu's name. She also asked her five sons to do the same since Viṣṇu alone is benevolent for all and provider of Mokṣa.

It is futile to initiate someone who does not understand the essence of the Pāhal Mantra, since to take this mantra is to make a vow to take refuge in Viṣṇu. One who does not understand this vow and resolution will treat this sacred water just as ordinary water. Without proper understanding, this ritual will not be able to transform the person. This water is a symbol of taking refuge in Viṣṇu so it is holy. Its characteristic is similar to water of river Ganga. It is as purifying and liberating as Ganga-water. If someone is initiated with this water by a saintly person with proper understanding, his entire sins are destroyed and he attains immense merits. What is this Pāhal? Let us try to understand the significance and essence of it. The 29 rules are that pond whose water is the pure character. To establish these 29 rules in one's mind firmly is to accept the Pāhal. Just hiding behind these rules is not enough but one has to practice them strictly. Firm mind is that king who washes away sand of all desires from his feet and remains established on the seat of 29 rules. Upliftment of creatures is possible only by following the path of the sages. Therefore, only well-deserved candidates should be given Pāhal who can remain well disciplined. Just as sandalwood and other fragrant trees are not present in every forest, lotus flowers are not present in every pond, not every god or goddess is capable of delivering mokṣa to creatures. Therefore, one should stop going around other gods and goddesses and only chant Viṣṇu so that all kinds of illusions and ignorance can be destroyed. It is futile to roam in 68 pilgrimage towns because there is none capable of delivering mokṣa other than Pāhal of Viṣṇu. Just as white cow is rarely found, just as a true sage is rarely found among saffron-robed people, similarly one who chants Viṣṇu continuously is the true devotee. This is the true significance of Pāhal. O devotees! A seeker who loves purity and character is the true noble sage. One who has taken this Pāhal of Viṣṇu, who has taken multiple incarnations, can be liberated from the cycle of life and death. Pāhal of the 10 incarnations of Viṣṇu is liberating.

Bālak Mantra

When a Bishnoi child is born, both the mother and the child are kept in a secluded room for 30 days. Both of them take complete rest in seclusion and are given nourishing food, the mother is not supposed to do any household chores for 30 days. Special attention is paid to the hygiene and silence in the room. People cannot visit a newborn baby for these 30 days. Cow-urine is kept in an open earthen pot that is supposed to protect the health of both mother and child from bacteria etc. It is ensured that direct sunlight, extreme heat, or extreme cold does not affect the baby or the mother. On the thirty-first day, the mother bathes the baby. The baby's hair is shaved and it is given new clothes. The entire house is cleaned and decorated. The Bishnoi priest *Gāyanāchārya* performs the fire-ritual, recites the śabdas of Guru Jambheśvara, establishes the Kalaśa with Pāhal-water, and then recites the Bālak-mantra:

> Om is the *Ādi-Brahman,* true guru, and *Niranjan.* The universe is created by the will of that Niranjan. The baby born in this house is also created by the creator of the universe Niranjan Brahman. By His grace and power, his power is manifested in this child. This child is also a form of Viṣṇu. Child's body is made of seven elements (*Rasa, Rakt, Mans, Asthi, Majja, and Sukr & Rajj*). For nine or ten months, fetus made of seven elements stayed in mother's womb, a dark dungeon like place, and grew there. In the womb, the baby was hanging upside down, i.e., head of the baby was below and legs were up. Only by God's grace, baby came out of this dark dungeon like place. Baby was cleaned with water and was removed of all impurities from its body with the water-deity. By hearing the name of Viṣṇu, his body and mind will be purified of all kinds of impurities. Thus, he will become clean and pure. As soon as the baby heard the sound of Viṣṇu-mantra and drank the Pāhal-water, he became a Bishnoi by the grace of Jambheśvara.

Guru Mantra

Although all Bishnois consider Jambheśvara as their spiritual guru, most Bishnois also regard another guru at the physical worldly level. This guru makes one disciple after the usual rituals of fire-ceremony, Kalaśa, water and recitation of śabdas. This ceremony, known as *Grahastha Mantra*, denotes one's entry into householder stage of life. It is usually performed at the wedding ceremony. One who does not have a guru is not preferred for marriage. This is the mantra recited as the guru mantra:

> Omkār is the true guru. The eternal subtle Brahman is manifested in the five elements but it is always one in the form of Atma. It is a yogi by its nature and dwells in the void. It has neither mother nor father. This *alakh-niranjan* is self-

> born into this world with five-element body. Ganga, Yamuna, and Saraswati of spiritual wisdom are flowing eternally but only rarely one takes a dip into it. Anybody can get liberation with this mantra and will not have to return to this world.

Wedding Ceremony

Bishnoi weddings are similar to other Indian weddings with minor differences. Firstly, the horoscopes of bride and groom are not matched. In addition, priests are not asked to find the auspicious time for the wedding. Astrological positionings are also not referred to, which are usually referred to for finding the auspicious timings for most Hindu weddings. Although remarriage is allowed for widow and widower, a widow's remarriage is less elaborate than that of a widower. For instance, a woman is not allowed to sit in the wedding seat for the second time but a man is. However, the social status of both remarriages is considered equal and the children born of such marriages are treated without any distinction or discrimination.

A Bishnoi wedding is different from other Hindu or Jain communities. Although they practice endogamy and leave the four gotras for marriages like other communities, they do not have astrological charts so there is no planetary matching involved of brides and bridegrooms like the common Hindu weddings. Moreover, their wedding rituals are much simpler. They do not have the circumambulation of fire seven times, which is performed in Hindu or Jain weddings. However, exchange of the seats of bride and bridegroom does take place; the bride comes from left to right after half of the ceremony, typically called *kanyādāna.* I was told that this ritual is based on the Śiva-Pārvatī wedding pattern. Dowry is not a common practice in Bishnoi weddings.

The levirate (remarrying the widow with the brother of the deceased husband), child marriage, and even polygamy are other peculiar characteristics of Bishnoi marriages (Fisher 1997: 72).

Dīkṣā Mantra

When a Bishnoi householder wishes to become a Bishnoi ascetic, he has to join the group of ascetics for about a year. He serves the senior monks and ascetics during this period. If he can adjust to the ascetic lifestyle during this period and if his guru approves of his new lifestyle, he is initiated into the Bishnoi monk order with a proper renunciation ceremony. On some auspicious day such as Amāvasyā, in the presence of monks and ascetics, the fire-ritual is performed with recitation of śabdas. Kalaśa is established with pāhal mantra. Then guru recites the *Dīkṣā Mantra* for renunciation and covers the new disciple with an ochre sheet that is called *bhesha-denā.*

After this ritual, the new disciple is given a new name replacing his older householder name. With this new name, he is assumed to take a new birth with his guru as his new father. Bishnoi sages are required to renounce their family lives completely. Unlike other Hindu priests, Bishnoi sages cannot be householders. Their main responsibilities are chanting the name of Viṣṇu, performing nightly rituals, *Jāgaraṇa*, at Bishnoi homes, performing fire-rituals, and performing wedding and pāhal rituals for Bishnoi householders. This guru-disciple tradition has continued for the last 500 years in the Bishnoi community. One guru can have several disciples that are called *guru-bhāis* (guru-brothers). After the death of the guru, generally his most senior disciple becomes the next guru. The *Dīkśā Mantra* is as follows:

> Om word is Brahman itself. One who chants quietly 'I am omkār myself' attains liberation with the help of this true word. He does not have to return to the path of life and death. Such liberated soul stays in the Viṣṇu-loka enjoying the company of Viṣṇu and does not suffer the old age. Therefore, Viṣṇu mantra is the only goal and support of this life. Whoever chants this mantra, will attain liberation. There is no difference between omkār and Viṣṇu. This quiet mantra is an epitome of cosmic soul.

Death Rites

Just as every major Bishnoi rite of passage is celebrated with elaborate rituals and mantras, death is also celebrated as a major event. Death songs are sung for someone dying at the age of 100 or beyond. Happiness is expressed at the arrival of the vehicle of the death-god. Unlike the 12-day period after death in other Hindu communities, Bishnois celebrate the mourning period only for three days. It is believed that the departed soul stays with its relatives only for three days after leaving the physical body and then it goes to other worlds. In these three days, the following rituals are performed:

a. The Last Bath – at the last breath, the person is removed from the deathbed and is placed on the floor. Then he or she is made to hear the kunchi-śabda of Jambheśvara. When the breathing and heartbeat are stopped and the dead body is still warm, it is bathed for the last time with pure water mixed with Ganga-water.
b. *Kafan* – after the last bath, the dead body is covered with the cotton clothes, called *kafan.* Men are clothed with white kafan, married women with red and unmarried women with black kafan. Ascetics are clothed with ochre kafan.
c. *Trin-Shaiyā* – after wearing kafan, a corpse-carriage is made with bundle of dry grass. The dead body is laid on this and the male relatives carry it to the cemetery with bare feet.

d. *Dāg* – around the death time, some relatives go to the cemetery and dig up a hole six-feet deep, six-feet long, and three-feet wide. The dead body is brought and buried inside this hole called *ghor*. The head of the body is kept northwards and all the relatives pour sand in the ghor and fill it completely. The filled ghor is then sprinkled with water. Lastly, the relatives take baths and return to their homes.
e. *Kāgol* – after coming home, the relatives, imagining that the deceased person is still present with them, prepare kāgol with his or her favorite food and feed it to crows. Kāgol is also given the next day. On the third day, some portion of the sweets made for the visiting guests is offered to the fire and then it is mixed with some yogurt. It is then taken to the cemetery and offered at the buried place, this is called the last kāgol. It is believed that the crows are the medium by which the food reaches the departed soul. Later, water is poured from an earthen pot on a khejari (or other green tree) through a woolen cloth having some grains on it. This is called *jalānjali*, offering of water. This is the final farewell given to the departed soul. It is believed that a Bishnoi reaches heaven after this last ritual. After giving the last kāgol and jalānjali, the relatives return home and drink three sips of pāhal water prepared by Bishnoi priest with mantras. This water is then sprinkled in the entire house and the sutak-period is completed. The relatives are then gifted with traditional clothes by the visiting guests to mark the end of their mourning period. This is the last "rite of passage" of Bishnois.

Appendix D
Translation of Jambheśvara Darśana by Brahmanand Sharma[1]

1. First to Jambheswar and then to the farmer I bow down. With the grace of guru Vidyādhara, I state the speech of Jambheśvara.
2. Where the jīva is explained, also the world is stated. The fundamental substance that liberates the jīva is enlightened here.
3. There are two activities considered in this speech: one is for the guru and the other is for the mind. By the former, one is liberated from the cycle of rebirth. By the latter, one is entangled in the cycle of rebirth.
4. The body ends so the death is inevitable. The dead is born again and thus the cycle of life and death continues.
5. The jīva should always strengthen and nourish the foundation and while doing that one can liberate from the bondage of life.
6. The supreme devotion is to concentrate one's mind on Vishnu and to chant Vishnu. The permanent peace is here.
7. Here three worlds and 14 abodes are considered. The divine light of the almighty enlightens them all.
8. The heaven is the abode of bliss and the hell is that of misery. The peace is present in the Vaikuntha where the liberated people reside.
9. There are several kinds of relationships in the world. There are different kinds of emotions. There are desired friends and relatives also.
10. These all perish, they do not have any essence so the wise person, having known the fundamental substance, does not experience happiness in them.
11. The home is not here but it is ahead. For how many days will the cows of the farmer stay?
12. There is only one home in the universe and that is in Vaikuntha. The incomplete practice is false. The fundamental condition stays only at one station.
13. On this land of the universe, there is no grain or water. O farmer! If you want to stay alive, remove your senses from here.
14. The guru is the supreme substance and is the abode of compassion. When guru is experienced directly, nothing else remains to be known.

---

1 This is my translation of a Sanskrit hymn eulogizing Jambheśvara written by Brahmanand Sharma (1986).

15. The supreme substance is devoid of the form, devoid of quality, and devoid of color. From its illusionary power, the universe arises from it.
16. This is also with quality, with shape, and is known as the Hari. This is the provider of peace and bliss and stays in the hearts of the devotees.
17. For the liberation of the people, at the call of the devotees, it comes to the earth in the human body.
18. The truth is the supreme dharma. Everything is established on the truth. The truth is the fundamental substance and is enjoined by it.
19. The shape is considered one whether internal or external. The practice should follow one's speech.
20. There are many covers on the universe. Therefore, the substance is covered. The people are roaming on the earth in several forms.
21. One should have compassion for all the creatures and should never harm anyone. Whether one lives or not, one should always protect other creatures carefully.
22. The caring for the cows is the supreme dharma. The Hari is the supreme example of it. Its milk is nourishing and its violence is prohibited.
23. The dharma of the farmer is the labor. One should earn the crops by hard work. This is a rightful earning that does not exploit anyone.
24. If one has more, one should give more. If one has less, one should give less. There is no scope for saying no. At least little should be given.
25. There are two kinds of purities in the world: physical and mental. The former is enjoined in the latter therefore bathing is to be done.
26. The milk of the cow should always be consumed. This increases the sattvik quality. Intoxicants are to be avoided. They increase the rajasic quality.
27. The action that purifies the mind should always be done with efforts. The fire ritual should be performed. It purifies everything.
28. The character is the supreme bathing. This is the biggest austerity. One, who has bathed in this, is considered to have bathed in nobility.
29. One who has controlled the sensual pleasures can earn rightfully and can cross the world by chanting Vishnu.

Appendix E
Athavale's Ecological Inspirations[1]

Athavale derived several inspirations from the BhP (11.7–9). This is a dialogue between Yadu and an avadhūta in which the avadhūta describes his 24 gurus as different objects of nature, the earth, air, sky, water, fire, moon, sun, pigeon, python, the ocean, moth, honeybee, elephant, the deer, the fish, the prostitute Pingala, the eagle, the child, the young girl, arrow maker, serpent, spider, and the wasp.

> Pṛthvi vāyur ākāśam, āpogniś candramā ravih, kapotojagarah sindhuh,
> Patango madhukṛd gajah, madhu-hā harino mînah, pingalā kurarorbhakah
> Kumāri śarakṛt sarpa, urnanābhih supeśakṛt, ete me guravo rājan
> Caturviṃśatir āśritāh, śikśā vṛttibhir eteṣām, anvaśikṣam ihātmanah.
> (11.7.33–5)

Earth – Earth is my first guru that I see first thing in the morning after waking up. Five great elements bother me but it continues to practice forgiveness. In Sanskrit, Earth is called *Sarvasahā*, one that bears all (Abbott 1974). Mountains and trees are also parts of the Earth, both of which are benevolent. Insane humans ramble on it, grow grains on it, dig out water from its belly, and yet forget to bow down to it. Such a human being cannot become spiritual seeker. Therefore, our ancestor has instructed us to bow down to the Earth while getting up in the morning, samudra vasane Devī. O mother Earth! You take care of me, I make my house on you, I walk on you, I produce grain from you, and I dig water from your belly. I am most related to you and that is why I adore you and bow to you. Mother! I torture you from my feet so please forgive me. When Brahmins establish a new idol in temple, they first worship the Earth. There is a deep significance for Earth-worship. Earth is forgiving and the rivers and trees born from its mountains are benevolent. Avadhūta says that he has learnt two qualities from the Earth, forgiveness and benevolence. One should be benevolent. "I am being benevolent" this should not occur to the donor. Similarly, a person accepting the favor should not feel too obliged for it. We should not harm one's self-respect by doing favor on him. Such charity is meaningless.

Wind – the second guru is wind: internal life force and external atmospheric wind. Internal life force breathes in oxygen but does not accumulate it since it does not have false sense of fear about shortage of oxygen. We accumulate water

[1] This section is based on my translation from Hindi discourses by Athavale published in Swadhyaya book *Vyāsa Vicāra*.

but not wind. Not accumulating is a higher stage of development and this I have learnt from wind, says Avadhūta. External wind brings the fragrance of flower and bad odor of gutter. Wind is detached from both fragrance and bad odor. Non-possessiveness and detachment, these two qualities I have learnt from the wind. We have to make our life beautiful by these two qualities. Who wants to reach God, should be detached. What do we do? We hold on to form and beauty and do not leave them at all. Secondly, air serves others but remains hidden. It does not advertise its service. We have to learn this quality from it. Avadhūta says that this quality is useful for life-development and I have learnt it from air.

Sky – the third guru is sky. Sky is a gift given to humans by nature. This is our laboratory, pleasure space, and health space. Just by sitting under the open sky will make our lives wealthy and enriched. If we observe the sky and its stars, we forget our sorrows spontaneously. If we observe it at night, we may not even need to sleep! The sky influences our minds due to its infiniteness. If we have to become a seeker, we have to learn to be infinite. This is why the Sanskrit word for sky is ananta, literally meaning infinite. Sky appears near us but it is quite far. Sky teaches us that our goal is not easy or near. As we go higher, the sky goes further away. Similarly, our goal in life should keep developing higher and higher. It is not easy to understand or realize God; we have to achieve this goal even if it takes multiple lifetimes. We have to develop this infinite perspective learning from sky. Sky is the roof of world-temple. Sun and Moon are hanging like chandeliers from it. This universe is created by God and is covered with the sheet of sky. Hence another Sanskrit word for sky is ambar, literally meaning cloth. This clothing is a token of grace from God. Sky keeps everything in its belly. If I have to develop, I will have to do the same. If I have to transform someone, I will have to cover that person. Sky is clean and untainted. When a storm comes, there are sand-particles, dry leaves, etc. everywhere. However, the sky becomes clean again later on. Birds also fly in the sky but sky remains detached from them. Vedanta has stated that God exists in our sky-heart. Heart is compared with the sky because when the heart becomes like sky, it will become divine. When a person is steady in all the opposite poles of life, he can realize and visualize God. Sky is also neutral to all. It accepts the sandstorm, the sun, the moon, and the birds but it does not belong to anyone. We should also accept diverse people in our lives without being influenced by them; this is the sign of a developed life. Maharashtrian saint Rāmdāsa Swāmi describes the beauty of Rāma by comparing it with sky, nabhāsārikhe rupa rāghava. Sky has no physical entity and it cannot be touched. Life should be like the sky. Sky is beyond time and it remains the same in all the eras and times. That which cannot be touched by time is sky. We also have to travel through three times, past, present and future. Past influences us, future inspires us, and present contains us. We have to transcend all three times. We should neither worry about the past, nor desire about the future, nor are attached to the present. If the present time is bad, we should not hate it. If it is good, we should not have affection for it. When we become as timeless as sky, our human life will become developed and prosperous. Therefore, we will have to adore the sky, and our hearts will become like sky and then God will reside in it.

Water – then the avadhūta says that he has learnt from water. Water is useful for rituals. We have to adore God. Out of five great elements, space, fire, wind, earth, and water, we can only utilize water in worshipping the God. Varuṇa, as a deity of water, is regarded very highly by priests in their rituals. Their rituals start with Varuṇa-Pūjā. Water is the easiest material to worship God. Water is clean, pure, and fluid. These are the qualities to be adopted to develop our lives. God is everywhere and is present in water so we bow to water. We forget our sorrows by looking at river Ganga. We like to offer not just coins but also our lives into its flow. Our lives appear small compared to Ganga. Our wealth, bank account, and property appear negligible in front of Ganga. We feel like throwing our materialistic lives into Ganga. A student of science describes water as H_2O; water has two components Hydrogen and Oxygen. However, Hydrogen gas is a combustible gas and Oxygen aids in combustion. Two such flammable gases result into the fire-resistant water, this is indeed a divine intervention! There is divinity in water. Perhaps this is why water of Ganga stays uncontaminated for several months. There is a Jala-Sūkta, water hymn, in the Vedas. Water is life. Water inspires talent. Ancient Vedic sages learnt from water and other natural resources, adored them and composed hymns about them. Nature is our most matured teacher. The Sanskrit word for water is paya, literally meaning nourishing element. Water sustains and nourishes life. It is an indispensable element for human life. Water also is neutral for all kinds of people, rich and poor, educated and uneducated, high and low, it does not discriminate among them. Water gives stimulus and enthusiasm to us. A touch of water on our eyes makes us fresh and awake from sleep. Water goes everywhere therefore it is egoless. It is lively and moving, not dead as stone. We do not regard still life as good. Even dharma is supposed to be dynamic but it has become still today. Life of a seeker should also be dynamic, neutral to all, enthusiastic, lively, says Avadhūta.

Fire – the fifth guru is fire. Fire is enlightening. Vedic culture regarded fire as the most supreme deity. Fire is to be worshipped first thing in the morning, "O Fire! Make me as illuminated as you. Make me victorious with your victorious glow. Make me destroyer of all impurities with your powerful flame." However, scriptures tell us that fire-worship will cease in Kaliyuga. Indeed, the fire is no longer worshipped in homes today, except in chain-smoking! Fire is illuminating, cold destroying, warmth giving, and invincible. It is also all witnessing. It is a silent witness at the time of wedding, havan, and death. Indian culture mentions the distinct mouths of God: Brahmin, the vocal mouth, and fire, the eating mouth. After death, our bodies are offered to the fire with the prayer that fire should take us to appropriate abode because it is the witness of all our karmas. This is the purpose of the *Antayeṣṭi Saṃskāra*, the final "rite of passage" performed by the son. Thus, fire has qualities of illumination, pain-destruction, non-possessiveness, purity, and warmth giving. A seeker should cultivate these qualities learning from fire. A Sanskrit word for fire is *udarabhājana*, one who only accepts something in its belly and does not accumulate. Fire does not just throw away everything but it does not accumulate them. Non-possessiveness is that quality which removes the insistence for accumulation. If we accept the wealth as a blessing from God and utilize it for divine work, it is non-possessiveness. Fire is uncontaminated and remains hidden. Just as fire is hidden inside wood, a seeker should also remain hidden and detached in the society.

Moon – the sixth guru is moon. Avadhūta told that he looked at the sky for light. One cannot look at the blazing Sun but can look at the cooling Moon so Moon is also guru. A seeker should regard Moon as guru. Indian children call him their maternal uncle, chandāmāmā. Just as Moon's phases increase and decrease, so should our life be. An ascending person should be mindful of his potential descent also. Moon teaches us that this body is impermanent. Just as there are 16 phases of Moon, our life also has 16 phases. We should orient them toward God. Just as Moon shines on a full-moon night with all its phases being completed, we will also become complete when all our 16 phases are completed. Phases of the Moon change in two fortnights, in the first one it earns its phases every night and in the next fortnight, it gives them away every night. On the night of the New Moon, it completely loses itself. Then again starts increasing in the next cycle. We should learn from the Moon how to earn laboriously and then donate it kindly and gracefully. This is a sign of self-respect and self-confidence. In the latter fortnight, Moon rises late in the night around 3am when the whole world is sleeping and no one is looking at him. It continues to do its work silently and secretly. During the latter part of our lives, we are not noticed or respected by anybody. We should learn from the Moon how to live peacefully even during our old age. This is what I learnt from the Moon, says Avadhūta.

Sun – Sun is the seventh guru. Indians have adored the Sun in the famous Gāyatri Mantra, tatsavitur vareṇyam ... Sun has qualities such as kindness, illumination, detachment, and selflessness. We adore the Sun to learn these qualities from him. We should also learn continuity and discipline from him. We should perform our duties according to our age. Just as everything is illuminated in the Sunlight, the light of knowledge should illuminate our minds. Just as there is no disturbance or pollution in the Sun, our minds should also be clear. The Sun is not just a fireball in the sky but it is the lifeline of every creature on the planet Earth. Life cannot even be imagined without the Sun. The ṚV says *surya ātma jagatastasthuśaśaca*; the Sun is the soul of every creature. It is psychological law that we become what we imagine and adore. Ancient Indians adored the Sun and tried to make their lives illuminating and meritorious. The Sun is also referred to as a friend of human beings and is adored in Sanskrit, mitrāya namah. The Sun brings light into our lives and removes sins born in darkness. Sunrise inspires us to start our actions. Thus, it removes lethargy from our lives. In addition to light, it also gives us heat and energy. Its heat maintains our body temperature at 98.4 degrees Fahrenheit. The Sun also accepts water from the Earth, makes it sweet, and returns to the Earth with the rain. The Sun is a great karmayogi and it is the ideal role model for other karmayogis. A true seeker perseveres on his chosen path with great efforts without complaining about lack of resources. A Sanskrit verse notes the lack of resources available to the Sun, "The Sun's chariot has only one wheel, its seven horses are reined by a snake, its path is without any support, and its charioteer Arun is without legs. Even with so many limitations, the Sun goes beyond the sky every day. The success lies in the true discipline, not in the availability of the resources." Indeed, the Sun rises at a precise time every day and makes time move forward. This teaches us a lesson of discipline. One who is disciplined never complains about lack of time. Sun also has great humility. It does not demand any salutation from anybody. It silently comes and waits at our doorsteps and enters our houses only if we open our doors or windows. It also does not need any name and fame for its work. Sun salutation is also a great yogic exercise that nourishes body and mind. It is recommended that we should observe both the dawn and the dusk everyday to illumine our minds and intellect.

Pigeon – the avadhūta considers pigeon his eighth guru. This bird teaches how excessive attraction for a thing can delude one's mind. A pigeon died when his whole family was caught by a hunter. His attachment to his family causes his own death also. We have to learn from the pigeon not to forget the self and the God. One should spend all one's energy for one's development and for God.

Python – the avadhūta says that the python is his ninth guru who taught him how to live without self-insistence. A python does not have his own insistence about his food. He just sits with his mouth open and eats whatever comes into his mouth. One should live with all his personal insistence and desires surrendered to God and accept only what is sent by God. One should live with indifference towards materialistic pleasures and should not run after them. Whatever is to be sent by God will definitely come our way so there is no need to spend our energy for that. We should rather utilize our energy for our development.

Sea – sea is the tenth guru of the avadhūta. Sea is traditionally considered important in Indian culture. Bathing in the sea is considered the best kind of bathing. Those who bath with the tap water, develop narrow mind. The water from well is better than tap water. Water from river is better than well. Water from confluence of two or more rivers is still better. Water from the sea is considered the best. Those who observe the sea everyday develop broad mind. Avadhūta says that the sea never leaves its boundary. It is happy externally but profound internally. A seeker should also be happy externally and profound internally. It is endless, unpolluted, calm, and incomprehensible. A seeker should learn these qualities from it. The BhG compares sthitaprajña, stable-mind, with the sea. There is both the high and the low tide in the sea but the sea remains stable in both. The sea gives us many gifts silently, e.g., rain, precious jewels, fish, etc.

Bug – the avadhūta's next guru is an insect that gets attracted by a candle and burns in it. Beauty attracts and instead of escaping from it, we should consider it divine.

Bee – the avadhūta's next guru is a bee that collects the essence of flowers. It collects the juice from several flowers instead of from one. We should not limit to a single book but should rather learn from different scriptures. Even an ascetic should not settle at one place but should keep roaming to avoid attachment to a place or a person.

Honeybee – the avadhūta's next guru is a honeybee. He learns the art of accumulation from them. The honeybee collects the honey but it is taken away by someone else. We should also accumulate the wealth for the sake of others. Wealth has three states, consumption, donation, or destruction. If we do not consume for ourselves or donate for others, it will be destroyed eventually.

Elephant – the avadhūta's next guru is a male elephant that desires to touch a female elephant. Even a powerful animal like the elephant falls due to its lust for the opposite sex. Lust should be replaced by sacred reverence for beauty. We should sublimate our desires not suppress them.

Deer – a deer could be caught by a hunter by pleasant music. An obscene art or obscene music is to be renounced. A seeker should not get obsessed by that art which obstructs one's development.

Fish – just as a fish is caught by angler by a bait, a person who is a slave to his senses remains a slave by materialistic pleasures. A seeker should rise beyond them.

Bird – a bird holding a piece of meat is bothered by others but when it throws away that piece, it is no longer bothered by others. This teaches Avadhūta the quality of non-possessiveness.

Pingalā – Pingalā was a prostitute who repents about misusing her body for lust. We should also utilize our body or mind not for lusts or desires but for God's work.

Arbhak – next, he meets a child who is happy without depending on any external object. We should also learn to be subjectively happy. A child also does not have desire for name and fame. Child's third quality is lack of anxiety and worry. A seeker should also renounce all desires for fame, and all kinds of anxieties and worries.

Young girl – once the avadhūta was roaming in the evening. He entered a home in which all the family members had gone out and only a young girl was present. Her marriage was already fixed and her future in-laws are visiting her at that time. She starts preparing food for them and to avoid the noise of her bangles, she breaks them all and leaves only one bangle in each hand. This teaches Avadhūta a lesson that one should remain alone to avoid unnecessary arguments and disturbances while meditating and worshipping.

Snake – the avadhūta tells that an ascetic should learn five things from a snake. First, he should live like a snake without building a permanent house. Second, he should avoid the crowd and live alone secretly like snake. Third, an ascetic should not accept spare things to accumulate, just as a snake does not even have hands to accept things. Fourth, he should practice his penance secretly just as a snake lives at secret places. Fifth, he should remain silent like a snake.

Arrow-maker – an arrow-maker concentrates on making arrows. This teaches Avadhūta the art and skill of focusing one's mind. External and internal noises should not distract a seeker from his goal or practice of meditation.

Insect – later, the avadhūta observes an insect shutting off its dwelling from another insect. Later it thinks about the invading insect and eventually becomes like it. This teaches Avadhūta the importance of thinking. One should think about sacredness that can change one's life.

Spider – a spider spins its web and, after some time, it swallows its own web. God also designs the universe with a thought and then dissolves it at the end of a cosmic cycle. This is the teaching that the avadhūta learns from a spider.

Athavale's Ecological Inspirations from Cultural Symbols

Athavale had also commented on several ecological elements present in Hindu rituals and myths.

Lotus – a Sanskrit verse says, "One which blooms in the water but whose fragrance spreads far away, whose trunk is hard, face is soft that shelters the friends, is addicted about cultivating good qualities, has animosity against bad qualities, goddess Lakṣami dwells in such Lotus blooming in the water, indeed she is integrated with Lotus." Among the various symbols of Indian culture, lotus holds a special place. Several gods and goddesses are depicted with lotus in their hands or even sitting on lotus. Maharashtrian saint Jñāneśvara has interpreted it in this way, "God wishes to worship the true devotee using Lotus." Indeed, this is a beautiful imagination. One whose life and character is as beautiful as lotus can earn the gift of lotus from God. In the BhG, Kṛṣṇa asks to live like lotus, "*Brahmaṇyādhāya karmāṇi saṇgam tyaktvā karoti yah. Lipyate na sa pāpena padmapatramivāmbhasā*" (BhG 5.10), meaning those who perform actions as sacrifice to God without attachment, sins cannot stick to them just as water does not stick to lotus. Lotus is a fine example of detachment. It teaches us how to live in the world and yet remain detached from its defects. Lotus does not complain about the mud and dirt around it but continues to develop beautifully even in its unclean surroundings. It always aims higher at the Sun. It blooms at the Sunrise and closes itself at the Sunset, thus its life is completely dependent on the Sun. If we also practice such disciplined devotion toward the Sun, we can also develop even in the dirty surroundings. Lotus is also a symbol of beauty. Poets have compared different parts of the body with lotus in Indian languages, for example, *hastakamal* (hands), *charaṇakamal* (feet), *hṛidayakamal* (heart), *nayanakamal* (eyes), *vadanakamal* (mouth), etc. In Indian architecture, sculpture, and literature, lotus is found at several places. Great poet Kālidāsa and great pundit Jagannanātha have also used lotus at several places in their literary works. Bhartṛhari even remarks that a pond without lotus pierces his heart like a needle. The beauty of water and lotus are mutually dependent. To summarize, lotus symbolizes ideal detachment, worship of Sun, creation of beauty, and philosophy of life. Perhaps this is why several gods and goddesses have accepted it as their seat, e.g., Brahma and Lakṣami.

Coconut – a Sanskrit verse says, "Noble people are like coconut while others are like berries that look attractive only externally." It is easier to look good externally but to become sweet and pleasant internally requires great efforts. Coconut teaches us to reform internally. A berry may look beautiful externally and may even taste better initially but its seed is hard and is eventually thrown away. On the other hand, coconut does not get embarrassed for its external ugliness and makes it soft and fair inside. One cannot control one's external beauty but one is capable to change and improve internally. Socrates and Abraham Lincoln are great examples who were known for the beauty of their thoughts and qualities. Great people are extremely strict in their self-discipline but are kind in disciplining others. Coconut is also used in rituals in the temples. Coconut resembles human head with its hair, tail, nose, and eyes. Perhaps earlier sacrifice of humans and animals was later replaced by offering coconuts in the rituals in ancient times. In this way, coconut helped human society develop from violence to non-violence.

Leaf – Śiva accepts Bilva leaf, Gaṇeśa accepts Durvā and Narayana accepts Tulasi. This signifies that even such small objects are sufficient to offer to gods. The reverential and devotional perspective in these offerings is important, not the material value of the object being offered. Leaf also means knowledge as mentioned in the BhG *chhandānsi yasya parṇāni*. This signifies that offering to God should be done with proper understanding and knowledge so that rituals do not become monotonous and mechanical.

Flower – offering flower signifies that the true offering to God is our actions, our speech, and our life. Flower has fragrance, colors, juice, and softness. Our life should also have fragrance of noble actions, colors of devotion, juice of knowledge, and softness of love.

Fruit – offering fruit signifies offering actions to God. Since our actions are performed with the power given to us by God, only God can be the true recipient of our actions. We have the right neither to accept the fruits of our action nor to renounce the action completely. Offering our actions and their results according to the teaching of the BhG shows us the middle path to be adopted. The results offered to God turns into divine blessing that brings happiness into our lives.

Vibhuti Yoga and Ecology – there can be three perspectives toward nature, consumption, appreciation, and reverence. Today, the most widely seen perspective is that of a consumer. Poetic appreciation can be seen in Kālidāsa's Sanskrit poetry and Romantic poetry of Europe. Like some other cultures, Indian culture teaches one to have the third perspective of reverence towards nature in addition to consumption and appreciation. For example, farmers should domesticate cows and bullocks not just for utility sake but also to practice love toward animal species. Cows and bullocks are the representatives of the animal kingdom and hence humans should love them. Similarly, *Tulasi* plant is to be revered as it represents botanical kingdom. With this perspective, Indian culture teaches nature worship. The Vibhuti Yoga of the BhG provides several examples of divinity in nature. This helps develop reverential perspective toward nature. Only this reverential perspective will make humans completely human and will make the entire universe divine. We will be able to regard everything as divine. The entire universe then turns into a teacher. Some humans regard fabricated books as divine but one who regards God-made books, i.e., natural resources as divine, is the real human. Ancient Indian sages studied nature and developed themselves. They learnt stability from Himalaya, Profundity from the sea, infiniteness from the sky, pacification from the moon, and tenderness from creepers and plants. Nature is a great teacher. There is a definite relationship between humans and other species. Therefore, Indians have worshipped trees. Our parents are to be revered beyond their financial support to us. Similarly, nature is to be revered beyond its consumer value for us. Reverential value includes four components, reverence for self, reverence for humans, reverence for animals, and reverence for all creation. In the tenth chapter Vibhuti Yoga, Kṛṣṇa mentions Vibhutis, divinities, in several species. In addition to consumption and appreciation, one should have gratitude, love, and reverence for nature. With this reverential attitude, humans will attain integral development and then the path of devotion starts. This is the experience of saints and seekers. Vibhuti Yoga is not merely a description of God but it also signifies and teaches reverence for all creation. Specifically, he mentions that following are some examples of his divine presence in different species. Among animals, he is Kāmadhenu among cows, Vāsuki among snakes, Uccaiśravas among horses, Airāvata among elephants, lion among wild animals, eagle among birds, and crocodile among fishes. Among natural entities, he is Himālaya among mountains, Meru among mountain-summits, Aśvattha among trees, ocean among water-reservoirs, and Ganga among rivers. He goes on to list several other examples where he is present and finally tells Arjuna that wherever he finds liveliness, wealth and beauty, he should regard that as divine. Thus, the Upaniṣadic principle *Iśāvāsyam idaṃ sarvam* is explained with specific examples.

Sāgar Pūjan (Ocean Worship)[2]

Like the trees and the cows, Athavale also included the ocean and the earth in his discourses to inspire his followers to develop reverence for them. In addition to reverence, the great qualities of the ocean and the earth can also help develop ethical qualities in humans, just as he sought in his previous prayogs of trees and cows.

On August 22, 2002, thousands of Swadhyayis converged at Girgaum, Chowpatty, Mumbai to celebrate the tenth anniversary of *Sāgar Pūjan*, a prayog

---

[2] Information in this section is based on my translations from Swadhyaya's Hindi book *Saṃskṛti Pūjan*.

for ocean worship started by Athavale in 1992 (*The Times of India*, September 30, 2002). Fisher folk from Gujarat, Maharashtra, and Goa witnessed the *Pūjan* that is traditionally performed on the full moon day of *Śrāvaṇ* month that also marks the beginning of the fishing season. Traders of communities such as Raghuvaṃśi, Luhanas, Thakkars, and Sindhis have been traditionally performing such rituals for several centuries. Historically, these communities have also been engaged in maritime activities for travel and trade.

Explaining the significance of Sāgar Pūjan, Athavale explained that the courageous merchants in the older days used to do their business and trade by ships and boats. They worship Varuṇa, the deity of the ocean, on this day to pray for pleasant weather conditions for smooth maritime trade. Lohana community accepts *Dariyālal* as their deity and reveres him throughout the year. The Sāgar Pūjan festival reminds one of ancient Indian trade routes to faraway countries including Egypt, Greece, China, and Indonesia. According to Athavale, Indian traders also propagated Indian art, culture, philosophy and other traditions to these remote countries. Sāgar Pūjan should remind Indians of these great achievements of their ancestors while offering coconut to the ocean during the Pūjan.

Sāgare Sarva Tīrthāṇi, all holy places are present at the ocean, according to this verse as cited by Athavale. According to an old Indian adage, the confluence, *saṇgam*, is more sacred than rivers, *saritā*, and *sāgar*, ocean, is more sacred than the *saṇgam*. He highlighted several qualities of ocean that can inspire and teach humans. Ocean accepts all the good waters of the pious rivers and the wastewater without complaints, just as BhG's *sthitaprajña* does not deviate from the bitter problems of life. In *Rāmacaritamānas*, the statements of Rāma are compared with ocean. Just as the ocean does not change in its dimension even after all the rivers end in it, one should not become proud and arrogant after getting wealth, name, and fame. Just as the ocean remains same in the ebbs and tides, one should always remain stable in happiness and sorrow. Just as the ocean donates its precious stones but does not publicize, one should donate secretly without publicizing. Just as one goes to the depth of the ocean to seek the pearls, one should go to the depth of the wisdom of saints. Just as ocean gets attracted to the moon, one should also get attracted to the saints and sages. Just as ocean offers its salty water to the sun that is transformed into pure water in the form of rain, one should offer oneself to the saints to become pure at heart.

List of Vṛkṣamandiras [from Anand (1999: 325)]

S. No.	Name	Address	Established on
1	Yājñavalkya	Kalawad Road, Rajkot	July 12, 1979
2	Valmiki	Maljeenjva, Veraval	Aug 8, 1980
3	Atri	Machhava, Mahesana	July 5, 1984
4	Bhargava	Fulsar, Bhavnagar	July 27, 1984
5	Vasishtha	Mathal, Kachchha	July 12, 1985
6	Kashyap	Mota Asrana, Bhavnagar	July 18, 1985
7	Patañjali	Ajod - Sokhada, Vadodara	July 26, 1985
8	Parashar	Shekhpar - Rajkot Highway, Surendranagar	July 11, 1986
9	Atri Kula	Machhava, Mahesana	Aug 1, 1985
10	Agastya	Bhagod, Valsad	June 22, 1987
11	Shandilya	Gangvav, Jamnagar	July 17, 1987
12	Gautama	Vanthali, Junagarh	July 7, 1989
13	Gautama Kula	Vanthali, Junagarh	-
14	Bhardwaj	Rampura, Bharuch	July 27, 1989
15	Shaunak	Rojhad Tekro, Sabarkantha	Aug 11, 1990
16	Jaimini	Devbhane, Dhulia, Mah.	July 12, 1991
17	Dadhichi	Adityana, Junagarh	July 21, 1993
18	Kaṇva	Ganpadar, Kachchha	July 12, 1994
19	Sandipani	Badgaon, Burhanpur, MP	June 16, 1997
20	Janak	Mudetha, Banaskantha	July 4, 1998
21	Vyas Darshan	Bamkheda, Dhule, Mah.	-
22	Dhaumya	Dharampur, Valsad	Aug 6, 1999
23	Kaushik	Kadarama, Surat	Aug 8, 1999
24	Sumantu	Kenedi, Jamnagar	Aug 9, 2000

Appendix F
History of Beneśvara, a Bhil Pilgrimage Center[1]

The Beneśvara site has one of the most famous pilgrimage centers for the Bhils. It is widely believed that the saint Māwajī Mahārājā performed austerities at this place in 1784. His followers today are mostly of the sāda samāj who sing rāsalīlā and āgalvāni. Māwajī Mahārājā is believed to have performed līlās at the Beneśvara "Dhāma" similar to Kṛṣṇa's Lilas at Gokul and Vrindavan and hence Beneśvara is sometimes referred to as the "Vrindavan of Bāgad." In Vikram Saṃvat 1771, Māwajī was born at Sabalapuri (Sābala). His father was Dālam Rishi and his mother was Kesar Bai. Māwajī is believed to be a miraculous child who performed several deeds that helped establish his spiritual authority among the villagers. His birth is believed to have been a result of the boon from a yogi. He wrote five texts in the cave of Dholagarh in his exile days that are dispersed in different temples today. The first book, Ratan Sāgar, is at Vishwakarma temple in Banswara, the second book, Sām Sāgar, is at Hari Mandir in Sābala, the third book, Megh Sāgar, is at Hari Mandir in Sheshapur, the fourth book, Prem Sāgar, is at Hari Mandir in Poonjapur, and the fifth book, Anant Sāgar, was stolen during the raids in Bājirav Peshwa. Some of the contemporary kings were Muhammad Shah in Delhi, Bājirav Peshwa in Pune, and Maharaval Ramsingh in Dungarpur. When Māwajī appeared as the saint, the king of Dungarpur was Maharaval Shivsingh and king of Udaipur was Maharana Sangramsingh. Māwajī never adopted any political role and his miraculous power alone helped him emerge as the spiritual leader of the masses. In his adolescent days, he chose Beneśvara as his place for performing the penance in Vikram Saṃvat 1784. Beneśvara Mahadev was his chosen deity, Iṣṭa Devatā. Māwajī Mahāraja performed his Rāsleela from the Māgha Shukla Purnima until the fifth day of the month at this site. Since the birth of Māwajī Mahārājā is in the Māgha month's shukla ekādaśī, his devout followers celebrate the fair at Beneśvara from Māgha shukla ekādaśī to panchami since vikram saṃvat 1797. In 2007 (vikram saṃvat 2063), the 266th anniversary of this fair was celebrated.

Māwajī Mahārājā married many times. His first marriage was with Rupabai, daughter of chhoti audichya of Kupada, the second one was with Vakhatbhai, daughter of Badi audichya of Gamada Bamaniya, the third one that took place "in a dream," was with a Gujarati patidar widow Manujī of Sagawada village, the

[1] This is my translation of a Hindi report published in *Rajasthan Patrika* by Devram Mehta (2007).

fourth one was with Sāhujī. The birthplace of Māwajī was Sābala, the "līlā" place was Beneśvara Dhāma, and the "karma" places were Punjapur, Lasada, Paloda, Dungarpur, Banswara, and the Dholagir Mountain. He spent the final part of his life at Sheshapur and passed away on Māgha Purnima in Vikram Saṃvat 1801. There is a temple also at his "nirvaṇa" place. Successive gurus have continued his lineage from 1714 CE onwards. After Māwajī served as the guru from 1714 to 1744, the leadership passed to Udiyananda Mahārājā from 1744 to 1773, and Mrs. Janakunwari from 1773 to 1855, who built the temples at Beneśvara, Paloda, and Sheshpur according to the Rajasthan government gazetteer. Because she had no children, she adopted Śivananda from Leelavāsa village who was installed as the next guru in 1855, after being qualified from Kashi. Śivananda served as the guru Peethadheesh until 1872. From his blessing, the king of Dungarpur Maharaval obtained a son and he built a temple at Sābala as a token of gratitude. The next guru Pīṭhādhīśa was Purnanandaji from 1872 to 1888, Kamalanandaji from 1888 to 1928, Paramanandaji from 1928 to 1944, and Devanandaji from 1945 to October 31, 1989. On November 11, 1989, Goswami Achyutanandaji Mahārājā was installed as the current Pīṭhādhīśa. Under his leadership, the Beneśvara fair begins with flag-hoisting at Beneśvara Dhāma.

Bibliography

Videos

Athavale, Pandurang Shastri. *Discourses in Hindi and Gujarati on the ṚV, the Upanishads, and the BhG*. Mumbai: Sat Vichar Darśana, 1990–2000.

Balmeãs, Thomas, Juliette Guigon, and Patrick Winocour. *Maharajah Burger*. New York: Filmakers Library, 2000.

Bazaz, Abir and Meena Gaur. *Swadhyaya*. New Delhi: Public Service Broadcasting Trust, 2003.

Benegal, Shyam. *Antarnaad: Inner Voice*. India: Suhetu Films, 1992.

Bishnoi, Gurvinder. *Jambheśvara*. Jodhpur: Vision 29G, 1990.

Devan, Abi, Sudhi Rajagopal, and Juan Carlos Ortiz. *Panihari the Water Women of India*. United States: Choices, Inc., 2006.

Gore, Albert, et al. *An Inconvenient Truth*. Hollywood: Paramount, 2006.

Oser, Marie. *The Amazing Bishnois of Rajasthan*, 2007.

Pike, Neil. *Appiko: To Embrace*, 2006.

Rao, B.V.P. *Willing to Sacrifice*. Visakhapatnam, India: Yamini Films, 1999.

Skiba, Malgorzata. *Eco Dharma*. New Delhi: Public Service Broadcasting Trust, 2006.

Books and Articles in Hindi/Rajasthani

Acharya, Krishnananda. *Śabdavāṇi Darśana*. Rishikesh: Bishnoi Mandir, 1993.

Acharya, Krishnananda. *Jambha Puraṇa*. Rishikesh: Bishnoi Mandir, 2003.

Acharya, Krishnananda. *Jambha Sāgara*. Rishikesh: Bishnoi Mandir, 2003.

Anand. *Darshan Mere Jharokhe Se*. Ahmadabad: Shraddha Prakashan, 1999.

Bhanavat, Mahendra. "Rājasthānī Janajīvana main Loka-MBh kī Gūnja," in *MBh in the Tribal and Folk Traditions of India*, ed. K.S. Singh. Shimla: Indian Institute of Advanced Study, 1993.

Bhīmapānúdúiyā. *Jāga jam̐bhaiyaā jām̐bho āvai*. Bīkānera: Sugunūī Prakāsīana, 1999.

Bishnoi, Kishnaram. *Guru Jambheśvara, Jīvana aur Sādhanā*. Hisar: Shri Guru Jambheśvara Publications, 2005.

Bishnoi, Krishan Lal. *Guru Jambhoji Avam Bishnoi Panth Ka Itihasa*. Sirsa: Sambhrathal Prakashan Abubshahar, 2000.

Bishnoi, Shrikrishna. *Bishnoi Dharma-Saṃskāra*. Bikaner: Dhok Dhora Prakashan, 1991.

Caturvedi, Paraśurāma. *Uttarī Bhārat Kī Santa Paramparā.* Second edition, Allahabad: Bhāratī Bhandār, 1964, pp. 332–7.

Eṣa Pantha Etat Karma. Mumbai: Sat Vichar Darśana Trust, 1999.

Gahlot, Jagadish Singh. *MāravārÚa Rājya kā itihāsa.* Jodhapura: Mahārājā Mānasimúha Pustaka Prakāsİa, 1991.

Jain, Rajesh. *Madhyakālīna Rājasthāna meṃ Jaina Dharma.* Vārāṇasī: Pārśvanātha Vidyāpitha, 1992.

Jain, Sagarmal and Vijay Kumar. *Sthānakwāsi Jain Paramparā kā Itihāsa.* Vārāṇasī: Pārśvanātha Vidyāpitha, 2003.

Jambha-Gītā.

Jambha Purāṇa by Krishnananda Acharya, 2003, Rishikesh.

Maheshwari, Hiralal. Jambheśvara, *Vishnoi Sampradāya Aura Sāhitya, Jambhavāni ke Pātha-Sampādana Sahita.* Calcutta: B.R. Publications, 1970.

Maheshwari, Hiralal. *Jambheśvara kī Sabadavāṇī, Mula Aur Tīkā.* Calcutta: B.R. Publications, 1976.

Maheshwari, Hiralal. *Jambheśvara.* New Delhi: Sāhitya Akademi, 1981.

Maheshwari, Hiralal. *Mehojī Krita Rāmāyaṇa.* Calcutta: Sat Sahitya Prakashan, 1984.

Mehta, Devram in *Rajasthan Patrika*, February 2, 2007.

Mishra, Anupam. *Aaj Bhi Khare Hain Talab.* Delhi: Gandhi Peace Foundation, 1993.

Nainsi, Munhata and Nārāyanúasimúha Bhātúī. *Māravārúa ra paraganm̄ú rī vigata.* Jodhapura: Rājasthāna Prācyavidyā Pratiṣṭhāna, 1968.

Ojha, Gaurishankar Hirachand. *Jodhpur Rājya Kā Itihāsa,* 2nd volume, 1941.

Padmanabha, P. and I.C. Srivastava. *Census of India, 1981. Series 18, Rajasthan.* Delhi: Controller of Publications, 1981.

Parik, Suryashankar. *Jāmbhojī kī Vāṇī, Jīvanī, Darśana, aur Hindi Artha Sahita Mūlavāṇī-pātha.* Bikaner: Vikāsa Prakāśana, 2001.

Rāma, Pemā. *Madhyakālīna Rājasthāna Memú Dhārmika Āndolana.* Rajasthan: Vanasthalī Vidyāpīṭha, Rājasthāna, 1977.

Rāma, Pemā (ed.). *Rājasthāna Meṃ Dharma, Sampradāya, Va, Āsthāeṁ.* Rajasthan: Vanasthalī Vidyāpīṭha, Rājasthāna, 2004.

Rāma, Tejā and Balabīra Śarmā. *Guru Jāmbhojī evaṃ Mahātmā Gāndhī kea Dharma-Darśana kā Samīkshātmaka Mūlyāṅkana.* Dillī: Nirmala Pablikeśans, 1996.

Śabda-vāṇi Darśana.

ŚABDA-vāni Jambha Sāgara.

Sahāraṇa, Shrikrishna. *Khejadali ke 363 Bishnoi Amar Shahīda.* Akhil Bharatiya Jeevaraksha Bishnoi Sabha.

Sahu, Banwarilal. *Bishnoi Loka Geet, Ek Sāṃskṛtik Mūlyāṅkana.* Bīkānera: Vikāsa Prakāśana, 2005.

Sanskṛti Pūjan. Mumbai: Sat Vichar Darśana Trust, 1997.

Sharma, Brahmanand. *Jambheśvara-darśana.* Jayapura: Brahmānanda Śarmā, 1986.

Smārikā.

Suthāra, Rohatāsa Kumāra. *Guru Jambhesīvara Vānúī: Dārsīanika Visīlesahanúa.* Kurukshetra: Nirmala Book Agency, 1994.

Tattwadeep. Mumbai: Sat Vichar Darśana Trust, 1997.
Vora, Rajiv. *Gandhi Marg*, Vol. 41, No. 2. March–April 1996.
Vyāsa Vicāra. Mumbai: Sat Vichar Darśana Trust, 1999.

Books and Journal Articles in English

Abbott, J. *The Keys of Power: A Study of Indian Ritual and Belief*. Secaucus, NJ: University Books, 1974.

Agrawal, Arun. *Environmentality: Technologies of Government and the Making of Subjects*. Durham, NC: Duke University Press, 2005.

Alley, Kelly. *On the Banks of the Ganga: When Wastewater Meets a Sacred River*. Ann Arbor: University of Michigan Press, 2002.

Altemeyer, Bob. "The Decline of Organized Religion in Western Civilization," *International Journal for the Psychology of Religion*, Vol. 14, No. 2, 2004, pp. 77–89.

Apffel-Marglin, Frederique and Purna Chandra Mishra. "Sacred Groves: Regenerating the Body, the Land, the Community," in *Global Ecology: A New Arena of Political Conflict*, ed. Wolfgang Sachs, pp. 197–207. London: Zed Books, 1993.

Apffel-Marglin, Frederique and Pramod Parajuli. "'Sacred Grove' and Ecology: Ritual and Science," in *Hinduism and Ecology: The Intersection of Earth, Sky, and Water*, ed. Christopher Key Chapple and Mary Evelyn Tucker, pp. 291–316. Cambridge, MA: Harvard University Press, 2000.

Athavale, Pandurang Shastri. *Hope of Humanity*. Mumbai: Sat Vichar Darśana Trust, 1997.

Athavale, Pandurang Shastri. *Light That Leads: Lectures Delivered by Rev. Pandurang Shastri Athavale in the United States*. Mumbai: Sat Vichar Darśana Trust, 1998a.

Athavale, Pandurang Shastri. *Glimpses of Life of Kṛṣṇa*. Mumbai: Sat Vichar Darśana Trust, 1998b.

Athavale, Pandurang Shastri. *Dawn of Divinity*. Mumbai: Sat Vichar Darśana Trust, 1999.

Athavale, Pandurang Shastri. *Radiant Rhapsody*. Mumbai: Sat Vichar Darśana Trust, 2000a.

Athavale, Pandurang Shastri. *The Systems, the Way, and the Work*. Mumbai: Sat Vichar Darśana Trust, 2000b.

Bakker, F.L. *The Struggle of the Hindu Balinese Intellectuals*. Amsterdam: VU University Press, 1993.

Balagangadhara, S.N. *The "Heathen in His Blindness ..." Asia, the West, and the Dynamic of Religion*. Leiden: Brill, 1994.

Barnhill, David Landis and Roger S. Gottlieb. *Deep Ecology and World Religions: New Essays on Sacred Ground*. Albany: State University of New York Press, 2001.

Bartley, C.J. *The Theology of Rāmānuja: Realism and Religion*. London: RoutledgeCurzon, 2002.

Baviskar, Amita. *In the Belly of the River, Tribal Conflicts over Development in the Narmada Valley*. New York: Oxford University Press, 1995.
Bergmann, Sigurd. Editorial, *Ecotheology*, Vol. 11.3, September 2006.
Bharucha, Rustom and Komal Kothari. *Rajasthan, an Oral History: Conversations with Komal Kothari*. New Delhi: Penguin Books, 2003.
Bhattacharyya, Swasti. *Magical Progeny, Modern Technology: A Hindu Bioethics of Assisted Reproductive Technology*. Albany: SUNY Press, 2006.
Biernacki, Loriliai. "Wilhelm Halbfass: India and Philology," *Religious Studies Review*, Vol. 33, No. 2, April 2007, pp. 95–102.
Bilimoria, Purushottama, Joseph Prabhu, and Renuka Sharma. *Indian Ethics*. Farnham, UK; Burlington, VT: Ashgate, 2007.
Bishnoi, Kishnaram and Narsiram Bishnoi. *Religion and Environment*. New Delhi: Commonwealth Publications, 2000.
Bishnoi, R.S. *A Blueprint for Environment: Conservation as Creed*. Dehra Dun: Surya Publication, 1992.
Brandis, Dietrich. *Indian Forestry*. Woking: Oriental Institute, 1897.
Brockmann, Herma and Renato Pichler. *Paving the Way for Peace: The Living Philosophies of Bishnois and Jains*. Delhi: Originals, 2004.
Bromley, David G. "Perspective: Whither New Religions Studies? Defining and Shaping a New Area of Study," *Nova Religio: The Journal of Alternative and Emergent Religions*, Vol. 8, Issue 2, 2004, pp. 83–97.
Brown, W. Norman. "Prakrit vanadava 'Tree Sap, Self-Control'," *Language*, Vol. 30, No. 1, 1954, pp. 43–6.
Bryant, Edwin F. and Maria L. Ekstrant. *The Hare Krishna Movement, The Postcharismatic Fate of a Religious Transplant*. New York: Columbia University Press, 2004.
Buitenen, van J.A.B. "Dharma and Mokṣa," *Philosophy East and West*, Vol. 7, 1957, p. 36.
Burghart, Richard. "The Founding of the Ramanandi Sect," *Ethnohistory*, Vol. 25, No. 2, 1978, pp. 121–39.
Cafaro, Philip and Monish Verma. "For Indian Wilderness," *Terra Nova: Nature and Culture*, Vol. 3, No. 3, 1998.
Callicott, J.B. and R. Ames (eds). *Nature in Asian Traditions of Thought: Essays in Environmental Philosophy*. Albany: SUNY Press, 1989.
Carter, Lewis F. *Charisma and Control in Rajneeshpuram: The Role of Shared Values in the Creation of a Community*. Cambridge: Cambridge University Press, 1990.
Chandla, M.S. *Jambheśvara: Messiah of the Thar Desert*. Chandigarh: Aurva Publications, 1998.
Chapple, Christopher Key. *Nonviolence to Animals, Earth, and Self in Asian Traditions*. Albany: SUNY Press, 1993.
Chapple, Christopher Key. "Jainism and Nonviolence," in *Subverting Hatred: The Challenge of Nonviolence in Religious Traditions*, ed. Daniel Smith-Christopher, pp. 13–24. Boston: Boston Research Center for the 21st Century, 1998a.

Chapple, Christopher Key. "Hinduism, Jainism, and Ecology," *Earth Ethics*, Vol. 10, No. 1, 1998b.

Chapple, Christopher Key. "Hinduism and Deep Ecology," in *Deep Ecology and World Religions*, ed. David Landis Barnhill and Roger S. Gottlieb, pp. 59–76. Albany: SUNY Press, 2001.

Chapple, Christopher Key. *Jainism and Ecology: Nonviolence in the Web of Life*. Cambridge, MA: Harvard University Press, 2002.

Chapple, Christopher Key and Mary Evelyn Tucker. *Hinduism and Ecology: The Intersection of Earth, Sky, and Water*. Cambridge, MA: Harvard University Press, 2000.

Ching, Julia. "Eastern Asian Religions," *World Religions, Eastern Tradition*, ed. Willard G. Oxtoby. Toronto: Oxford University Press, 1996.

Clarke, J.J. *Oriental Enlightenment: The Encounter between Asian and Western Thought*. London: Routledge, 1997.

Cort, John. "Green Jainism? Notes and Queries toward a Possible Jain Environment Ethic," in *Jainism and Ecology: Nonviolence in the Web of Life*, ed. Christopher Key Chapple. Cambridge, MA: Harvard University Press, 2002.

Crawford, S. Cromwell. *Dilemmas of Life and Death: Hindu Ethics in North American Context*. Albany: State University of New York Press, 1995.

Creel, Austin B. "Dharma as an Ethical Category Relating to Freedom and Responsibility," *Philosophy East and West*, Vol. 22, No. 2, On Dharma and Li, 1972, pp. 155–68.

Creel, Austin B. "The Reexamination of 'Dharma' in Hindu Ethics," *Philosophy East and West*, Vol. 25, No. 2, 1975, pp. 161–73.

Creel, Austin B. *Dharma in Hindu Ethics*. Calcutta: Firma KLM, 1977.

Dasgupta, Shashi Bhushan. *Obscure Religious Cults*. Calcutta: Firma K.L. Mukhopadhyay, 1969.

Deutsch, E. "Vedanta and Ecology," in *Indian Philosophical Annual* 7, ed. T.M.P. Mahādeva. Madras, The Centre for Advanced Study in Philosophy, 1970.

Deutsch, E. "A Metaphysical Grounding for Natural Reverence, East-West,", in *Nature in Asian Traditions of Thought: Essays in Environmental Philosophy*, ed. J.B. Callicott and R. Ames. Albany: SUNY Press, 1989.

Dhand, Arti. "The Dharma of Ethics, the Ethics of Dharma: Quizzing the Ideals of Hinduism," *Journal of Religious Ethics*, Vol. 30, No. 3, Fall 2002 , pp. 347–72(26).

Dharampal-Frick, Gita. "Swadhyaya and the 'Stream' of Religious Revitalization," in *Charisma and Canon: Essays on the Religious History of the Indian Subcontinent*, ed. Vasudha Dalamia, Angelika Malinar, and Martin Christof. New Delhi: Oxford University Press, 2001.

Doniger, Wendy, J. Duncan, and M. Derrett (eds). *The Concept of Duty in South Asia*. Columbia: South Asia Books for the School of Oriental and Africa Studies, 1978.

Dundas, Paul. *The Jains*. London: Routledge, 2002 [1992].

Dwivedi, O.P. "Dharmic Ecology," in *Hinduism and Ecology: The Intersection of Earth, Sky, and Water*, ed. Christopher Key Chapple and Mary Evelyn Tucker. Cambridge, MA: Harvard University Press, 2000.

Dwivedi, O.P. and Lucy Reid. "Women and the Sacred Earth: Hindu and Christian Ecofeminist Perspectives," *Worldviews* 11, 2007, pp. 305–23.

Edgerton, Franklin. "Dominant Ideas in the Formation of Indian Culture," *Journal of the American Oriental Society*, Vol. 62, No. 3, 1942, pp. 151–6.

Elwin, Verrier. *The Religion of an Indian Tribe*. [Bombay]: Oxford University Press, 1955.

Feldhaus, Anne. *Water and Womanhood: Religious Meanings of Rivers in Maharashtra*. New York: Oxford University Press, 1995.

Fisher, R.J. *If Rain doesn't Come: An Anthropological Study of Drought and Human Ecology in Western Rajasthan*. New Delhi: Manohar, 1997.

Fitzgerald, Timothy. *Ideology of Religious Studies*. New York: Oxford University Press, 2000.

Fitzgerald, Timothy. "Problems with 'Religion' as a Category for Understanding Hinduism," in *Defining Hinduism: A Reader*, ed. J.E. Llewellyn. New York: Routledge, 2005.

Flügel, Peter. "Review of *Jainism and Ecology: Non-Violence in the Web of Life*, edited by Christopher Key Chapple," *Environmental Ethics*, Vol. 27, No. 2, 2005, p. 201.

Freeman, J. Rich. "Folk-Models of the Forest Environment in Highland Malabar," in *The Social Construction of Indian Forests*, ed. Roger Jeffery. New Delhi: Manohar Publishers, 1998.

Freeman, J. Rich. "Gods, Groves, and the Culture of Nature in Kerala," *Modern Asian Studies*, Vol. 33, No. 2, May 1999.

Fukuoka, Masanobu. *The One-Straw Revolution*. Emmaus, Pennsylvania: Rodale Press, 1978.

Gaeffke, Peter. "Karma in North Indian Bhakti Traditions," *Journal of the American Oriental Society*, Vol. 105, No. 2, 1985, pp. 265–75.

Gahlot, Jagadish Singh. *Māravārúa Rājya kā itihāsa*. Jodhapura: Mahārājā Mānasimúha Pustaka Prakāsía, 1991.

Gaur, Mahesh and Hemlata Gaur. "Combating Desertification: Building on Traditional Knowledge Systems of the Thar Desert Communities," *Environmental Monitoring and Assessment*, Vol. 99, No. 1–3, 2004, pp. 1–3.

Gier, Nicholas F. "Hindu Titanism," *Philosophy East and West*, Vol. 45, No. 1, 1995, pp. 73–96.

Gier, Nicholas F. "Toward a Hindu Virtue Ethics," in *Contemporary Issues in Constructive Dharma*, Volume 2, ed. R.D. Sherma and A. Deepak, pp. 151–162. Hampton, VA: Deepak Heritage Books, 2005.

Gier, Nicholas F. "A Response to Shyam Ranganathan's Review of *The Virtue of Non-Violence: From Gautama to Gandhi*," *Philosophy East and West*, Vol. 57, No. 4, October 2007, pp. 561–3.

Gillan, Michael. "Bengal's Past and Present: Hindu Nationalist Contestations of History and Regional Identity," *Contemporary South Asia*, Vol. 12, No. 3, 2003.

Giri, Ananta Kumar. "Review of Thomas Pantham Political Theories and Social Reconstruction, Survey of the Literature on India," *Gandhi Marg*, Vol. 17, No. 2, 1995, pp. 233–5.

Giri, Ananta Kumar. *Self-Development and Social Transformations? The Vision and Practice of the Self-Study Mobilization of Swadhyaya.* Lanham: Lexington Books, 2009.

Glucklich, Ariel. *The Sense of Adharma.* New York: Oxford University Press, 1994.

Gold, Ann G. "Of Gods, Trees and Boundaries, Divine Conservation in Rajasthan," *Asian Folklore Studies*, Vol. 48, 1989.

Gold, Ann G. "Sin and Rain, Moral Ecology in Rural North India," in *Purifying the Earthly Body of God: Religion and Ecology in Hindu India*, ed. Lance Nelson. Albany: SUNY Press, 1998.

Gold, Ann G. "Story, Ritual, and Environment in Rajasthan," in *Sacred Landscapes and Cultural Politics: Planting a Tree*, ed. Philip P. Arnold and Ann Grodzins Gold. Farnham, UK: Ashgate, 2001.

Gold, Ann G. "Malaji's Hill, Divine Sanction, Community Action," *Context: Built, Living and Natural*, Vol. 3, No. 1, 2006, pp. 33–42.

Gold, Ann G. and Bhoju Ram Gujar. *In the Time of Trees and Sorrows: Nature, Power, and Memory in Rajasthan.* Durham, NC: Duke University Press, 2002.

Gold, Daniel. *The Lord as Guru: Hindi Sants in North Indian Tradition.* New York: Oxford University Press, 1987.

Goldman, Robert P. *The Rāmāyaṇa of Vālmiki: An Epic of Ancient India. Bālakānda.* Princeton: Princeton University Press, 1990.

Gosling, David L. *Religion and Ecology in Indian and Southeast Asia.* New York: Routledge, 2001.

Greenough, Paul R. and Anna Lowenhaupt Tsing (eds). *Nature in the Global South: Environmental Projects in South and Southeast Asia.* Durham, NC: Duke University Press, 2003.

Greenwold, Stephen M. "Monkhood versus Priesthood in Newar Buddhism," in *Contributions to the Anthropology of Nepal*, ed. Christoph von Furer-Haimendorf, pp. 129–49. Guildford: Biddles Limited, 1974.

Gruzalski, Bart. "Gandhi's Contributions to Environmental Thought and Action," *Environmental Ethics*, Vol. 24, 2002, pp. 227–42.

Guha, Ramachandra. "Radical American Environmentalism and Wilderness Preservation: A Third World Critique," in *Ethical Perspectives on Environmental Issues in India*, ed. George A. James. New Delhi: APH Publishing Corporation, 1999.

Guha, Ramachandra. *The Unquiet Woods: Ecological Change and Peasant Resistance in the Himalaya.* Berkeley: University of California, 2000.

Guha, Ramachandra. *How Much Should a Person Consume? Environmentalism in India and the United States.* Berkeley: University of California, 2006.

Guha, Ramachandra and Madhav Gadgil. *This Fissured Land: An Ecological History of India.* New York: Oxford University Press, 1992.

Guha, Sumit. *Environment and Ethnicities in India: 1200–1991*. Cambridge: Cambridge University Press, 2006.

Gupta, Bina. "Bhagavad Gītā as Duty and Virtue Ethics: Some Reflections," *Journal of Religious Ethics*, Vol. 34, No. 3, 2006, pp. 373–95.

Gupta, S.C. *Mukam: A Village Survey. Census of India, 1961, Vol XIV, Rajasthan. Part VI-A, Village Survey No, 4*, 1965.

Haberman, David L. *Journey through the Twelve Forests: An Encounter with Krishna*. New York: Oxford University Press, 1994.

Haberman, David L. *River of Love in an Age of Pollution: The Yamuna River of Northern India*. Berkeley: University of California Press, 2006.

Hacker, Paul. "Dharma in Hinduism," *Journal of Indian Philosophy*, Vol. 34, 2006, pp. 479–96.

Halbfass, Wilhelm. *India and Europe: An Essay in Understanding*. Albany: SUNY Press, 1988.

Halbfass, Wilhelm. *Tradition and Reflection: Explorations in Indian Thought*. Albany: SUNY Press, 1991.

Halbfass, Wilhelm. *Philology and Confrontation: Paul Hacker on Traditional and Modern Vedanta*. Albany: SUNY Press, 1994.

Hardacre, Helen. *Kurozumikyo and the New Religions of Japan*. Princeton: Princeton University, 1988.

Hargrove, Eugene C. "Foreword," in *Nature in Asian Traditions of Thought: Essays in Environmental Philosophy*, ed. J. Baird Callicott and Roger T. Ames. Albany: SUNY Press, 1989.

Hawley, J.S. and Mark Juergensmeyer, trans. *Songs of the Saints of India*. New York: Oxford University Press, 1988.

Haynes, Edward S. "Natural and the Raj: Customary State Systems and Environmental Management in Pre-Integration Rajasthan and Gujarat," in *Nature and the Orient: The Environmental History of South and Southeast Asia*, ed. Richard H. Grove, Vinita Damodaran and Satpal Sangwan. New Delhi: Oxford University Press, 1998.

Holdrege, Barbara A. "Dharma," in *The Hindu World*, ed. Sushil Mittal and Gene R. Thursby. New York: Routledge, 2004.

Holtz, Barry W. *Back to the Sources: Reading the Classic Jewish Texts*. New York: Summit Books, 1984.

Ikeuchi, Fuki and Alison Freund. "Japanese Buddhist Hospice and Shunko Tashiro," in *Buddhist-Christian Studies*, Vol. 15, 1995, pp. 61–5.

Inden, Ronald. *Imagining India*. Bloomington: Indiana University Press, 2000.

Ingalls, Daniel H.H. "Dharma and Mokṣa," *Philosophy East and West*, Vol. 7, No. 1/2, 1957, pp. 41–8.

Islam, M. Zafar-ul and Asad R. Rahmani. *Human Influence on Biodiversity Conservation of the Thar Desert of India*. Mumbai: Bombay Natural History Society, 2002.

Jain, A., Katewa, S.S., Galav, P.K. and Sharma, P. "Medicinal Plant Diversity of Sitamata Wildlife Sanctuary, Rajasthan, India," *Journal of Ethnopharmacology*, Vol. 105, No. 2, 2005.

Jain, Pankaj. "Householders and Renouncers, The Holistic Combination in Indian Thought," in *Studies in Vedanta: Essays in Honor of S.S. Rama Rao Pappu*, edited by P. George Victor and V.V.S. Saibaba, pp. 165–80. New Delhi: D.K. Printworld, 2006.

Jain, Prem Suman and Raj Mal Lodha (eds). *Medieval Jainism, Culture and Environment*. New Delhi: Ashish Publishing House, 1990.

Jaini, Padmanabh S. *The Jaina Path of Purification*. Berkeley: University of California Press, 1979.

James, George A. (ed.). *Ethical Perspectives on Environmental Issues in India*. New Delhi: APH Publishing Corporation, 1999.

James, George A. "Ethical and Religious Dimensions of Chipko Resistance," in *Hinduism and Ecology: The Intersection of Earth, Sky, and Water*, ed. Christopher Key Chapple and Mary Evelyn Tucker. Cambridge, MA: Harvard University Press, 2000.

James, George A. "The Environment and Environmental Movements in Hinduism," in *Contemporary Hinduism: Ritual, Culture, and Practice*, ed. Robin Rinehart. Santa Barbara: ABC-CLIO, 2004.

James, George A. "Athavale" and "Swadhyaya," in *The Encyclopedia of Religion and Nature*, ed. Bron Taylor. New York: Thoemmes Continuum, 2005.

Jeffery, Roger (ed.). *The Social Construction of Indian Forests*. New Delhi: Manohar Publishers, 1998.

Jha, M., H. Vardhan, S. Chatterjee, K. Kumar, and A.R.K. Sastry. "Status of orans (Sacred Groves) in Peepasar and khejarli Villages in Rajasthan," in *Conserving the Sacred, for Biodiversity Management*, ed. P.S. Ramakrishnan, K.G. Saxena, and U.M. Chandrashekhara. Enfield: Science Publishers, 1998.

Johnston, Douglas. *Faith-Based Diplomacy: Trumping Realpolitik*. New York: Oxford University Press, 2003.

Joshi, Lal Mani. *Studies in the Buddhistic Culture of India, during the 7th and 8th Centuries A.D.* Delhi: Motilal Banarsidass, 1967.

Joshi, Prabhakar. *Ethnobotany of the Primitive Tribes in Rajasthan*. Jaipur: Printwell, 1995.

Journal of Indian Philosophy, Vol. 32, 2004.

Kalam, M.A. "Sacred Groves in Coorg, Karnataka," in *The Social Construction of Indian Forests*, ed. Roger Jeffery, pp. 39–54. New Delhi: Manohar Publishers, 1998.

Kane, Pandurang Vaman. *History of Dharmaśāstra*, Volume 2, Part 1. Poona: Bhandarkar Oriental Research Institute, 1974, pp. 1–11.

Kaplan, Cora, Debra Keats, and Joan Wallach Scott. *Transitions, Environments, Translations: Feminisms in International Politics*. New York: Routledge, 1997.

Kassam, Tazim R. *Songs of Wisdom and Circles of Dance: Hymns of the Satpanth Isma'ili Muslim Saint, Pir Shams*. Albany: SUNY Press, 1995.

Khan, Dominique-Sila. *Conversions and Shifting Identities: Ramdev Pir and the Ismailis in Rajasthan*. New Delhi: Manohar Publishers & Distributors: Centre de Sciences Humaines, 1997a.

Khan, Dominique-Sila. "The Coming of Nikalank Avatar: A Messianic Theme In Some Sectarian Traditions of North-Western India," *Journal of Indian Philosophy*, Vol. 25, 1997b, pp. 401–26.

Khan, Dominique-Sila. *Crossing the Threshold: Understanding Religious Identities in South Asia*. London: I.B. Tauris & Co. Ltd., 2004.

Khan, Dominique-Sila and Zawahir Moir. "The Lord Will Marry the Virgin Earth: Songs of the Time to Come," *Journal of Indian Philosophy*, Vol. 28, 2000, pp. 99–115.

Koller, John M. *The Indian Way: An Introduction to the Philosophies and Religions of India*. Upper Saddle River: Prentice Hall, 2005.

Korom, Frank J. "Holy Cow! The Apotheosis of Zebu, or Why the Cow is Sacred in Hinduism," *Asian Folklore Studies*, Vol. 59, No. 2, 2000, pp. 181–203.

Kumar, Nalini, Naresh Saxena, Yoginder Alagh, and Kinsuk Mitra. *India: Alleviating Poverty through Forest Development*. Washington, DC: World Bank, 2000.

Lal, Vinay. "Too Deep for Deep Ecology," in *Hinduism and Ecology: The Intersection of Earth, Sky, and Water*, ed. Christopher Key Chapple and Mary Evelyn Tucker. Cambridge, MA: Harvard University Press, 2000.

Lal, Vinay. "Bishnoi," in *The Encyclopedia of Religion and Nature*, ed. Bron Taylor. London: Continuum, 2005.

Landry, Elizabeth. *An Introduction to the Fifteenth Century Rajasthani Saint Jāmbho and His Vāṇī*. A Thesis for Master of Arts, University of Washington, 1990.

Larson, Gerald. "'Conceptual Resources' in South Asia for 'Environmental Ethics'," in *Nature in Asian Traditions of Thought: Essays in Environmental Philosophy*, ed. J. Baird Callicott and Roger T. Ames. Albany: SUNY Press, 1989.

Larson, Gerald. "'A Beautiful Sunset ... Mistaken for Dawn', Some Reflections on Religious Studies, India Studies, and the Modern University," *Journal of American Academy of Religion*, Vol. 72, No. 4, 2004.

Levitt, Peggy. "Redefining the Boundaries of Belonging: the Institutional Character of Transnational Religious Life," *Sociology of Religion*, Spring 2004.

Little, John T. "Video Vachana, Swadhyaya and Sacred Tapes," in *Media and the Transformation of Religion in South Asia*, ed. Laurence A. Babb and Susan S. Wadley, pp. 254–81. Philadelphia: University of Pennsylvania Press, 1995.

Lodrick, Deryck O. *Sacred Cows, Sacred Places: Origins and Survivals of Animal Homes in India*. Berkeley: University of California Press, 1981.

Lorenzen, David N. "Who Invented Hinduism?" *Comparative Studies in Society and History*, Vol. 41, No. 4, 1999, pp. 630–59.

Lorenzen, David N. "Bhakti," in *The Hindu World*, ed. Sushil Mittal and Gene R. Thursby. London: Routledge, 2004.

Lutgendorf, Philip. *The Life of a Text: Performing the Rāmcaritmānas of Tulsidas*. Berkeley: University of California Press, 1991.

Maathai, Wangari. "The Cracked Mirror," *Resurgence Magazine*, Issue 227, November 11, 2004.

Madan, T.N. *Non-Renunciation: Themes and Interpretations of Hindu Culture*. New York: Oxford University Press, 1996.

Mādhava and Tapasyananda. *Śaṅkara-Dig-Vijaya: The Traditional Life of Sri Śaṅkarācārya*. Madras: Sri Ramakrishna Math, 1978.

Majumdar, R.C., Hemchandra Raychaudhuri, and Kalikinkar Datta. *An Advanced History of India*. London: Macmillan, 1949.

Malhotra, Kailash C., Yogesh Gokhale, Sudipto Chatterjee, and Sanjiv Srivastava. *Sacred Groves in India: An Overview*. New Delhi: Aryan Books International, 2007.

Mann, R.S. *Nature, Man, Spirit Complex in Tribal India*. New Delhi: Concepts Publishing, 1981.

Mann, R.S. *Culture and Integration of Indian Tribes*. New Delhi: M.D. Publications Pvt Ltd, 1993.

Mann, H.S. and S.K. Saxena. *khejri (Prosopis Cineraria) in the Indian Desert: Its Role in Agroforestry*. CAZRI monograph, no. 11. Jodhpur, India: Central Arid Zone Research Institute, 1980.

Manuel, Peter. *Cassette Culture: Popular Music and Technology in North India*. Chicago: University of Chicago Press, 1993.

Marriott, McKim. "Little Communities in an Indigenous Civilization,", in *Village India*. Chicago: University of Chicago Press, 1965.

Marriott, McKim. "Interpreting Indian Society: A Monistic Alternative to Dumont's Dualism," *The Journal of Asian Studies*, Vol. 36, No. 1, 1976, pp. 189–95.

Marriott, McKim (ed.). *India Through Hindu Categories*. New Delhi: Sage Publications, 1990.

Marriott, McKim and Ronald Inden. "Caste Systems," *Encyclopedia Britannica*, fifteenth edn Chicago: Encyclopedia Britannica, 1981, Macropaedia III, p. 983.

Martin, Nancy M. and Joseph Runzo (eds). *Ethics in the World Religions*. Oxford: Oneworld, 2001.

Matilal, Bimal Krishna. *Ethics and Epics*, edited by Jonardon Ganeri. Oxford and New York: Oxford University Press, 2002.

McCall, Grant. *Dharma Dynamic: A Strategy for Anthropology in the Indian Village*. New Delhi: Cosmo, 1982 [1987] [2003].

McNeely, J.A., K.R. Miller, W.V. Reid, R.A. Mittermeier, and T.B. Werner. *Conserving the World's Biological Diversity*. International Union for Conservation of Nature and Natural Resources/World Resources Institute/Conservation International/World Wildlife Fund/US. World Bank, Gland, 1990.

Melton, Gordon J. "An Introduction to New Religions," in *The Oxford Handbook of New Religious Movements*, ed. James R. Lewis. New York: Oxford University Press, 2004.

Meyer, Marvin and Kurt Bergel (eds). *Reverence for Life: the Ethics of Albert Schweitzer for the Twenty-First Century*. Syracuse: Syracuse University Press, 2002.

Michaels, Axel and Barbara Harshav. *Hinduism: Past and Present*. Princeton: Princeton University Press, 2004.

Miller, Timothy. *When Prophets Die: The Postcharismatic Fate of New Religious Movements*. Albany: SUNY Press, 1991.

Mohanty, J.N. "Dharma, Imperatives and Tradition: Toward an Indian Theory of Moral Action," in *Indian Ethics: Classical Traditions and Contemporary Challenges*, ed. Purushottama Bilimoria, Joseph Prabhu and Renuka Sharma. Farnham, UK: Ashgate, 2007.

Mumme, Patricia Y. "Models and Images for a Vaiṣṇava Environmental Theology, The Potential Contribution of Śrīvaisnavism," in *Purifying the Earthly Body of God: Religion and Ecology in Hindu India*, ed. Lance Nelson. Albany: SUNY Press, 1998.

Nagarajan, Vijaya. "Embedded Ecology," in *Purifying the Earthly Body of God: Religion and Ecology in Hindu India*, ed. Lance Nelson. Albany: SUNY Press, 1998.

Nagarajan, Vijaya. "Rituals of Embedded Ecologies, Drawing Kolams, Marrying Trees, and Generating Auspiciousness," in *Hinduism and Ecology: The Intersection of Earth, Sky, and Water*, ed. Christopher Key Chapple and Mary Evelyn Tucker. Cambridge, MA: Harvard University Press, 2000.

Nanda, Meera. *Dharmic Ecology and the Neo-Pagan International: The Dangers of Religious Environmentalism in India.* Paper presented at the eighteenth European Conference on Modern South Asian Studies. Sweden: Lunds University, 2004.

Narayanan, Vasudha. "'One Tree is Equal to Ten Sons': Hindu Responses to the Problems of Ecology, Population and Consumption," in *Journal of American Academy of Religion*, Vol. 65, No. 2, 1997.

Narayanan, Vasudha. "Hindu Ethics and Dharma," in *Ethics in the World Religions*, ed. Joseph Runzo and Nancy M. Martin. Oxford: Oneworld, 2001.

Nelson, Lance. *Purifying the Earthly Body of God: Religion and Ecology in Hindu India.* Albany: SUNY Press, 1998.

Nugteren, Albertina. *Belief, Bounty, and Beauty: Rituals around Sacred Trees in India.* Leiden and Boston: Brill, 2005.

Olivelle, Patrick. "Semantic History of Dharma the Middle and Late Vedic Periods," in *Journal of Indian Philosophy*, Vol. 32, No. 5–6, December 2004

Orr, Matthew. "Environmental Decline and the Rise of Religion," *Journal of Religion and Science*, Vol. 38, Issue 4, December 2003.

Pande, Govindchandra. *Life and Thought of Śaṅkarācārya*. New Delhi: Motilal Banarsidass, 1994.

Pandey, Deep N. *Beyond Vanishing Woods: Participatory Survival Options for Wildlife, Forests and People*. Udaipur: Himanshu Publications, 1996.

Pandey, Deep N. *Ethnoforestry: Local Knowledge for Sustainable Forestry and Livelihood Security*. Udaipur: Himanshu Publications, 1998.

Parajuli, Pramod. "No Nature Apart: Adivasi Cosmovision and Ecological Discourses in Jharkhand, India," in *Sacred Landscapes and Cultural Politics: Planting a Tree*, ed. Philip P. Arnold and Ann Grodzins Gold. Farnham, UK: Ashgate, 2001.

Paranjape, Makarand. *Dharma and Development: The Future of Survival.* Delhi: Samvad India Foundation, 2005.

Parel, Anthony. *Gandhi's Philosophy and the Quest for Harmony*. Cambridge: Cambridge University Press, 2006.

Peritore, Patrick N. "Environmental Attitudes of Indian Elites, Challenging Western Postmodernist Models," *Asian Survey*, Vol. 33, No. 8, August 1993, pp. 804–18.
Potter, Karl H. *Presuppositions of India's Philosophies.* Delhi: Motilal Banarsidass Publishers, 1991.
Ramakrishnan, P.S., K.G. Saxena, and U.M. Chandrashekara (eds). *Conserving the Sacred for Biodiversity Management.* New Delhi: Oxford and IBH, 1998.
Robbins, Paul. *Political Ecology: A Critical Introduction.* Malden: Blackwell, 2004.
Rose, Deborah Lee and Birgitta Saflund. *The People Who Hugged the Trees: An Environmental Folk Tale.* Niwot: R. Rinehart, 1990.
Rukmani, T.S. *Turmoil, Hope, and the Swadhyaya.* Montreal: CASA Conference, 1999.
Sale, Kirkpatrick. "The Forest for the Trees: Can Today's Environmentalists tell the Difference," *Mother Jones*, Vol. 11, No. 8 (November 1986): 26.
Sanford, A. Whitney. "Shifting the Center: Yakṣas on the Margins of Contemporary Practice," *Journal of the American Academy of Religion*, Vol. 73, No. 1, December 2005, pp. 89–110.
Sanford, A. Whitney. "Pinned on Karma Rock: Whitewater Kayaking as Religious Experience," *Journal of the American Academy of Religion*, Vol. 75, No. 4, December 2007, pp. 875–95.
Sankhla, K.S. and Peter Jackson. "People, Trees and Antelopes in the Indian Desert," in *Culture and Conservation: The Human Dimension in Environmental Planning*, ed. Jeffrey A. McNeely and David Pitt. London: Croom Helm, 1985.
Sawhney, Aparna. *The New Face of Environmental Management in India.* Farnham, UK; Burlington, VT: Ashgate, 2004.
Saxena, N.C. *World Bank and Forestry in India.* India Country Study Background Paper. Operations Evaluation Department (OED). Washington, DC: World Bank, 1999.
Schomer, Karine and W.H. McLeod (eds). *The Sants: Studies in a Devotional Tradition of India.* Berkeley: Berkeley Religious Studies Series; Delhi: Motilal Banarsidass, 1987.
Sebastian, Sunny. "Rajsamand Lake: Death by Mindless Mining," in *The Hindu Survey of the Environment.* Madras: Kasturi & Sons Ltd, 2006.
Sessions, George. *Deep Ecology for the Twenty-First Century.* Boston: Shambhala Publications, 1994.
Sethi, Nitin and Ramya Viswanath. "A Profane Proposal," *Down to Earth*, December 31, 2003.
Shah, Tushaar. *Management of Natural Resources.* Mumbai: Sir Dorabji Tata Trust, 2003.
Shah, V.P., N.R. Sheth, and Pravin M. Visaria. "Swadhyaya: Social Change through Spirituality," in *Social Change Through Voluntary Action*, ed. M.L. Dantwala, Harsh Sethi, and Pravin M. Visaria. New Delhi: Sage Publications, 1998.
Sharma, Arvind. A Metaphysical Foundation for the Swadhyaya Movement (If It Needs One). Unpublished paper, 1999.

Sharma, Arvind. *Classical Hindu Thought: An Introduction.* New Delhi: Oxford University Press, 2000.

Sharma, Arvind. *The Study of Hinduism.* Columbia: University of South Carolina Press, 2003.

Sharma, Arvind. "'Religion' and 'Religious Freedom' Towards an Indic Understanding," in *Evam* (3/1), ed. Makarand Paranjape. New Delhi: Samvad India Foundation, 2003.

Sharma, Devinder. "Shedding Tears over Failed Watersheds," *Business Line*, June 14, 2000.

Sharma, Sudhirendar. *Food Security through Rainwater Catchment.* New Delhi: UNDP – World Bank Water Sanitation Program, 1999.

Sheth, N.R. "A Spiritual Approach to Social Transformation," in *The Other Gujarat*, ed. Takashi Shinoda. Mumbai: Popular Prakashan, 2002.

Shiva, Vandana. *Staying Alive: Women, Ecology and Development.* London: Zed Books, 1988.

Shiva, Vandana. *Stolen Harvest: The Hijacking of the Global Food Supply.* Cambridge, MA: South End Press, 2000.

Shresth, Swasti and Shridhar Devidas. *Forest Revival and Traditional Water Harvesting: Community Based Conservation at Bhaonta-Kolyala, Rajasthan, India.* Kalpavriksh, New Delhi and IIED, London, 1999.

Singer, Milton. *When a Great Tradition Modernizes: An Anthropological Approach to Indian Civilization.* New York: Praeger, 1972.

Singh, Aman. *"Oran" Sustainable Livelihood & Biodiversity Conservation System in Rajasthan Experiences from KRAPAVIS (Kṛṣi Avam Paristhitiki Vikas Sansthan)*, n.d.

Singh, Aman. "Oran – A Traditional Biodiversity Management System in Rajasthan," *Magazine on Low External Input and Sustainable Agriculture LEISA*, Vol. 5, No. 3, September 2003.

Singh, Aman. "Influencing Policy For Revival of oran Lands," *Magazine on Low External Input and Sustainable Agriculture LEISA*, Vol. 7, No. 4, December 2005.

Singh, D.K. "Lost Tribes," *Communalism Combat*, Vol. 11, No. 102, 2004.

Singh, G.S. and K.G. Saxena. "Sacred Groves in the Rural Landscape, A Case Study of Shekhala Village in Rajasthan," in *Conserving the Sacred, for Biodiversity Management*, ed. P.S. Ramakrishnan, K.G. Saxena, and U.M. Chandrashekhara. Enfield: Science Publishers, 1998.

Singh, K. Suresh and Birendranatha Datta. *Rama-Katha in Tribal and Folk Traditions of India: Proceedings of a Seminar.* Calcutta: Anthropological Survey of India, 1993.

Sinha, Ashish, Kevin G. Cannariato, Lowell D. Stott, Hai Cheng, R. Lawrence Edwards, Madhusudan G. Yadava, R. Ramesh, and Indra B. Singh. "A 900-year (600 to 1500 A.D.) Record of the Indian Summer Monsoon Precipitation from the Core Monsoon Zone of India," *Geophysical Research Letters*, Vol. 34, Issue 16, 2007, p. L16707.

Sinha, B.K. "The Answers Within," *Down to Earth*, May 1998.

Smart, Ninian and Shivesh Thakur. *Ethical and Political Dilemmas of Modern India.* New York: St. Martin's Press, 1993.

Smith, Frederick M. "The Hierarchy of Philosophical Systems According to Vallabhācārya," *Journal of Indian Philosophy*, Vol. 33, No. 4, August 2005.

Smith, Frederick M. *The Self Possessed: Deity and Spirit Possession in South Asian Literature and Civilization.* New York: Columbia University Press, 2006.

Sontheimer, Gunther-Dietz and Hermann Kulke (eds). *Hinduism Reconsidered.* New Delhi: Manohar, 1997 [1989].

Srinivas, Tulasi. "Divine Enterprise: Hindu Priests and Ritual Change in Neighborhood Hindu Temples in Bangalore," *Journal of South Asian Studies*, Vol. 29, Issue 3, December 2006, pp. 321–43.

Srivastava, R.K. *Vital Connections: Self, Society, God. Perspectives on Swadhyaya.* New York: Weatherhill Publications, 1998.

Staal, Frits. "Sanskrit and Sanskritization," *Journal of Asian Studies*, Vol. 12, No. 3, May 1963, pp. 261–75.

Tagore, Rabindranath. *Sādhanā: The Realisation of Life.* New York: The Macmillan Company, 1916.

Thapar, Romila. "The Householder and the Renouncer in the Brahmanical and Buddhist Traditions," in *Way of Life, King, Householder, and Renouncer*, ed. T.N. Madan. New Delhi: Vikas, 1982.

Thapar, Valmik. *Land of the Tiger, A Natural History of the Indian Subcontinent.* Berkeley: University of California Press, 1997.

Tilak, Srinivas. *The Myth of Sarvodaya: A Study of Vinoba's Concept.* New Delhi: Breakthrough, 1984.

Tobias, Michael. *Life Force: The World of Jainism.* Fremont: Asian Humanities Press, 1991.

Tod, James. *Annals and Antiquities of Rajasthan, or, the Central and Western Rajpoot States of India.* London: G. Routledge & Sons, 1914.

Tomalin, Emma. "The Limits of Religious Environmentalism for India," *Worldviews: Environment, Culture, Religion*, Vol. 6, No. 1, 2002, pp. 12–30.

Tomalin, Emma. "Bio-Divinity and Bio-Diversity: Perspectives on Religion and Environmental Conservation in India," *Numen*, Vol. 51, No. 3, 2004, pp. 265–95.

Tomalin, Emma. *Bio-Divinity and Bio-Diversity: The Limits to Religious Environmentalism.* Farnham: Ashgate, 2009.

Tyndale, Wendy. *Visions of Development: Faith-Based Initiatives.* Farnham, UK: Ashgate, 2006.

Unterberger, Betty Miller and Rekha R. Sharma. "Shri Pandurang Vaijnath Athavale Shastri and the Swadhyaya Movement in India," *Journal of Third World Studies*, Vol. 7, No. 1, Spring 1990, pp. 116–32.

Vallely, Anne. "From Liberation to Ecology: Ethical Discourses among Orthodox and Diaspora Jains," in *Jainism and Ecology: Nonviolence in the Web of Life*, ed. Christopher Key Chapple. Cambridge, MA: Harvard University Press, 2002a.

Vallely, Anne. *Guardians of the Transcendent: Ethnography of a Jain Ascetic Community.* Toronto and Buffalo: University of Toronto Press, 2002b.

Van Horn, Gavin. "Hindu Traditions and Nature: Survey Article," *Worldviews: Environment, Culture, Religion*, Vol. 10, No. 1, 2006, pp. 5–39(35).

Vasan, Sudha and Sanjay Kumar. "Situating Conserving Communities in their Place: Political Economy of Kullu Devban," *Conservation and Society*, Vol. 4, No. 2, June 2006, pp. 325–46.

Vertovec, Steven. *The Hindu Diaspora: Comparative Patterns*. London: Routledge, 2000.

Vitebsky, Piers. "A Farewell to Ancestors? Deforestation and the Changing Spiritual Environment of the Sora," in *Nature and the Orient: The Environmental History of South and Southeast Asia*, ed. Richard H. Grove, Vinita Damodaran, and Satpal Sangwan, pp. 967–82. New Delhi: Oxford University Press, 1998.

Vohra, Ashok, Arvind Sharma, and Mrinal Miri. *Dharma: The Categorical Imperative*. Delhi: Sundeep Prakashan, 2005.

Waldau, Paul and Kimberley C. Patton. *A Communion of Subjects: Animals in Religion, Science, and Ethics*. New York: Columbia University Press, 2006.

Weber, Thomas. *Hugging the Trees: The Story of the Chipko Movement*. New Delhi: Penguin Books, 1990.

Weightman, Simon and S.M. Pandey. "The Semantic Fields of Dharma and Kartavy in Modern Hindi," in *The Concept of Duty in South Asia*, ed. Wendy Doniger, J. Duncan, and M. Derrett. Columbia, MO: South Asia Books for the School of Oriental and Africa Studies, 1978.

White, Lynn, Jr., "The Historical Roots of Our Ecological Crisis," *Science*, Vol. 155, No. 3767, March 1967.

Wilson, Bryan R. *Religion in Sociological Perspective*. Oxford: Oxford University Press, 1982.

Wilson, Bryan R. *The Social Dimension of Sectarianism: Sects and New Religious Movements in Contemporary Society*. New York: Oxford University Press, 1990.

Index

Acārāṅga Sūtra 100, 101
Advaita 11, 25, 58, 110, 123
Agarwal, Anil 48
Alley, Kelly 8
Amrita Devi 10, 51, 63, 65, 66
Appiko Movement 10, 34, 65
Ascetic Model – *See* Renouncer Model
Athavale, Pandurang Shastri Athavale 10, 11, 17–49

Bahuguna, Sunderlal 10, 11, 17, 52, 98
Balagangadhara, S. N. 127
Bhatt, Chandiprasad 10
Bhattacharyya, Swasti 124–5
BhG (Bhagvad Gītā) 10, 19–22, 29, 33, 35, 40, 42, 108, 126, 145, 165, 167, 185–190
Bhil xi, 1, 2, 10, 14, 15, 76, 79–94
BhP (Bhāgavata Purāṇa) 10, 40, 42, 181
Bible 5, 6, 29
Bishnoi xi, 9, 10–15, 40, 46, 49, 51–77
Blackbuck 49, 57, 67, 69–73, 130

Callicott, Baird 122
Chapple, Christopher 8, 12, 48, 100, 112, 117, 118
Chatterjee, Bankim Chandra 106
Chipko Movement 10, 11, 34, 52, 65, 76, 130
Christianity 5, 14, 25, 26, 27, 28, 83, 95, 96, 103
Consumerism 1, 9, 28
Cow 39, 86, 98, 99, 100, 123, 135–8, 144–5, 148–9, 150, 155–8, 160, 163, 166–7, 174–5, 179, 180, 188–9
Creel, Austin 110–112

Deep ecology 8, 10–12, 32, 75–6, 127–8
Delhi 43, 58, 71, 99, 193
Desert 9, 51–3, 57, 62–3, 73–6, 118–9, 126–8, 131, 142–3, 147–8, 152, 155–8, 161, 163–4, 169
Devotional Model 6–9, 117–8
Dhand, Arti 108–9, 115, 120
Dharma xi, 1, 4, 105–116
Doniger, Wendy 58, 106–7, 114
Down to Earth 9, 68, 73, 82, 89
Durgā 85, 88
Dwivedi, O. P. 52, 76, 112

Earth ix, 8, 18, 31, 34, 39, 41–9
Embedded Ecologies 119, 124
Environmental ethics 3, 4, 6, 31, 33, 41, 115, 117, 120, 122–4
Ethnosociology 3, 109, 118

Feldhaus, Anne 6, 7, 96, 109
Fisher, R. J. 66, 111–12, 119, 176
Fitzgerald, Timothy 115

Gadgil, Madhav 15
Gaeffke, Peter 119
Ganga (Ganges) 7, 8, 61, 98, 123, 161, 162, 174, 176, 177, 183, 189
Gandhi, Mohandas 8, 10, 25, 28, 29, 30, 77, 102, 106, 112, 118–9, 127
Gier, Nicholas 110, 111, 113
Glucklich, Ariel 3, 7, 113–5
Gold, Ann 46, 52, 76, 80, 82, 93, 109, 119, 120, 127
GreenFaith 95–6
Guha, Ramachandra 13, 52, 63, 75–6, 84, 93, 102, 121, 124, 127, 128, 130–1
Guha, Sumit 83
Gujarat 17–49
Gupta, Bina 124

Haberman, David 110, 117–18, 120, 127
Halbfass, Wilhelm 25, 95–6, 105–6, 120, 127
Hanumān 73, 74, 85, 87, 90, 149, 151, 152, 157, 165, 169

Hargrove, Eugene 122
Hawley, Jack 10, 56
Himalaya 94, 123, 162, 189
Hindustan Times 69

ISKCON 2
Islam 14, 25–7, 55–6

Jainism 28, 101, 103
Jains 8, 99, 101, 117, 129
Jala-Sūkta 183
Jambheśvara, Guru 9, 10, 40, 51–77
James, George 10, 112, 120
Jodhpur 9, 49, 51–3, 58, 61–75

Kabir 56, 58, 83
Khan, Dominique-Sila 55
Khejari 9, 51–77
Kṛṣṇa 17, 19, 21, 26, 32, 38, 40, 42, 57, 83, 85, 90, 108, 135, 138–9, 142, 145–6, 148–9, 152–6, 160–1, 165–7, 171, 187–9, 193

Larson, Gerald 10, 30, 105, 118
Lodrick, Deryck 99
Lutgendorf, Philip 108

Madhya Pradesh 19, 23, 31, 53, 84
Maharashtra 7, 9, 19, 22–48
Manu Smṛti 7, 74, 108
Marriott, McKim 3, 30, 84, 105, 112, 118
Matilal, Bimal 110
MBh (Mahābhārata) 7, 22, 74, 77, 84, 108–9, 111, 124, 149, 165–6
McCall, Grant 112–5
Mohanty, J. N. 107–8
Mumbai 17, 19, 20, 24, 27, 29, 189
Muslims 29, 39, 45, 55–6, 59, 60, 108, 137, 151, 162–3

Nanak 56, 58
Narayanan, Vasudha 6, 7, 8, 12, 105, 109, 112, 115, 117, 120, 125
Neem 53, 81, 85, 94, 138, 140
Nelson, Lance 2, 6, 9, 11, 112, 120
New Religions, New Religious Movements 19–31
Nirmal Nīrs 41–46
Non-violence 8, 56, 60, 100, 107, 129, 188

Ocean Worship 189–190

Pīpal 32, 53, 62, 81, 85, 89, 94
Prahlād 52, 56, 149, 160, 163, 164, 173, 174
Purāṇa 7, 10, 11
Puruṣa sūkta 106, 107

Rajasthan 1, 2, 9, 10, 15, 49–94, 194
Rāma 33, 34, 60, 90–91, 109, 142, 144, 151–3, 157, 160, 165–170
Rāmānuja 7, 11
Rāmāyaṇa 6, 7, 33, 74, 77, 84, 90, 109, 111, 165, 168–170
Renouncer Model 6–10, 117–118
Rukmani, T. S. 18, 26, 111, 112

Sacred Groves 2, 9, 13, 14, 38, 72, 73, 79–94
Sanford, Whitney 25, 85, 95, 102, 105, 111
Śaṅkara 6, 11, 12, 110, 111
Sharma, Arvind 107
Shallow Ecology 32, 127
Śiva 26, 32, 33, 44, 74, 85–8, 149, 150, 171, 176, 188
Smith, Frederick 7, 14, 26, 85, 109, 123, 130
Śrisūktam 38, 42
Swadhyaya xi, 1,2, 3, 9, 10, 11, 12, 13, 14, 15, 17–49

The Hindu 66, 75, 99
Times of India 29, 36, 67–8, 71, 75, 131, 190
Tobias, Michael 99
Tomalin, Emma 1, 12–15, 96, 112
Tree-temple 17, 31
Tulsi 32, 33, 56
Tulsidas 56, 168

UK 17, 21, 38, 100, 121
Upaniṣada 7, 19, 32, 34, 40
USA 19, 21, 26, 38, 75, 100, 131

Vallabha 11, 110–111, 117
Vallely, Ann 101, 117
Vaiśnavism 11
Vedanta 6, 9, 110–2, 182

Vegan 100
Vegetarianism 9, 83, 98–100, 118–122, 129
Vertovec, Steven 85
Viṣṇu 25, 29, 41–44, 52–58, 83, 137–177
Vishnoi 52
White, Lynn 5, 13

Yamuna 48, 162, 176
Yogeśvara Kṛṣi 46–48